Thinking in Images

Also available from Bloomsbury:

Adorning Bodies, by Marilynn Johnson
Aesthetics, Philosophy and Martin Creed, edited by Elisabeth Schellekens and Davide Dal Sasso
Experimental Philosophy of Identity and the Self, edited by Kevin Tobia
Human Beings and their Images, by Christoph Wulf
The Philosophy and Science of Predictive Processing, edited by Dina Mendonça, Manuel Curado and Steven S. Gouveia

Thinking in Images

Imagistic Cognition and Non-propositional Content

Piotr Kozak

BLOOMSBURY ACADEMIC
LONDON · NEW YORK · OXFORD · NEW DELHI · SYDNEY

BLOOMSBURY ACADEMIC
Bloomsbury Publishing Plc
50 Bedford Square, London, WC1B 3DP, UK
1385 Broadway, New York, NY 10018, USA
29 Earlsfort Terrace, Dublin 2, Ireland

BLOOMSBURY, BLOOMSBURY ACADEMIC and the Diana logo are trademarks of
Bloomsbury Publishing Plc

First published in Great Britain 2023
This paperback edition published 2024

Copyright © Piotr Kozak, 2023

Piotr Kozak has asserted his right under the Copyright, Designs and
Patents Act, 1988, to be identified as Author of this work.

For legal purposes the Acknowledgements on p. ix constitute an extension
of this copyright page.

Cover image: Suprematism No. 56, 1916 Kazimir Malevich 1878–1935 Russia,
USSR, Russian, Federation, © Peter Horree / Alamy Stock Photo

All rights reserved. No part of this publication may be reproduced or transmitted
in any form or by any means, electronic or mechanical, including photocopying,
recording, or any information storage or retrieval system, without prior
permission in writing from the publishers.

Bloomsbury Publishing Plc does not have any control over, or responsibility for, any
third-party websites referred to or in this book. All internet addresses given in this
book were correct at the time of going to press. The author and publisher regret any
inconvenience caused if addresses have changed or sites have ceased to
exist, but can accept no responsibility for any such changes.

A catalogue record for this book is available from the British Library.

A catalog record for this book is available from the Library of Congress.

ISBN: HB: 978-1-3502-6746-6
PB: 978-1-3502-6750-3
ePDF: 978-1-3502-6747-3
eBook: 978-1-3502-6748-0

Typeset by Jones Ltd, London

To find out more about our authors and books visit www.bloomsbury.com and
sign up for our newsletters.

To Kasia, Tadeusz and Józef

Contents

List of Figures	viii
Acknowledgements	ix
Introduction	1
1 What is the problem of thinking with images?	17
2 What is thinking?	27
3 What answers should we expect?	55
4 What do images do?	75
5 Recognition-based identification	101
6 What is an image?	123
7 Thinking with images	149
8 Conclusion	181
Notes	185
Literature	196
Index	225

Figures

2.1	Theories of thinking	50
4.1	The unknot (on the left) and an equivalent knot (on the right)	76
4.2	Reidemeister moves	77
4.3	The left-trefoil knot and the right-trefoil knot	78
4.4	A square pattern matched with a tangram set	80
4.5	The construction of the triangle ΔFGK	82
4.6	Line graph representing inflation growth	88
4.7	The pulley system diagram	88
4.8	Line graphs (a) and bar charts (b) convey the same information but in a different way, affecting the accessibility of the information	92
4.9	Euler diagrams representing the subset, intersection and disjoint relations, respectively	93
4.10	Two diagrams of a triangle with the same construction rules	97
4.11	Depending on the identified rules of construction, ΔABD and ΔACE are similar or congruent to each other	97
5.1	The first picture of a black hole	102
5.2	The cross-ratio invariant	113
5.3	Munker-White's Illusion	115
5.4	Images impose order on some manifold to localize the elements of the structure	118
6.1	The 2-D model of iconic reference representing the relationship between an image, its content, the referent and the target	125
6.2	The relation between trueness, precision and accuracy conditions	134
6.3	Reutersvaard-Penrose triangle	143
6.4	The impossible fork	145

Acknowledgements

Books never have a single author. They are always the fruit of discussion. I would like to thank Bartosz Działoszyński, Paweł Gładziejewski, Witold Hensel, Mateusz Hohol, Małgorzata Koronkiewicz, Paweł Kozak, Mira Marcinów, Jakub Matyja, Bence Nanay, Kristóf Nyíri, Michał Piekarski, Robert Poczobut, Robert Rogoziecki and Nastazja Stoch for inspiring discussions and valuable advice. I am very grateful for the anonymous reviewers for their helpful remarks. Special thanks to Maja Białek, Krystyna Bielecka, Marcin Miłkowski and Marek Pokropski, who read the early version of the manuscript for their insightful comments. I am also grateful to Kamil Lemanek and Krzysztof Gajda for proofreading.

I wrote this book during the Covid-19 pandemic and finished it when the Russo-Ukrainian War started. It was not a good time to write books. The only explanation why this book was written is the continuous support of my family. Thus, I would like to express my deepest gratitude to the most thoughtful and caring person I know, who happily is my wife, and the people I love the most, who happen to be my children.

Chapter 2 includes material from my paper 'The Diagram Problem', *Diagrammatic Representation and Inference 2020*. LNCS (pp. 217–24). Springer.

Chapter 6 includes parts of my paper 'The Analog-Digital Distinction Fails to Explain the Perception-Thought Distinction: An Alternative Account of the Format of Mental Representation', *Studia Semiotyczne*, 35 (1), 73–94.

This work was supported by the research grant 'What Is Thinking with Images?', SONATA 10, granted by the National Science Centre, Poland, based on the decision No. 2015/19/D/HS1/02426.

Introduction

Let me begin with a philosophical confession of faith on what a philosophical problem is. The nature of a philosophical problem is ultimately bound up in knowledge and wonder: granted that knowledge begins with wonder, the nature of wonder seems to differ between science and philosophy, respectively. In science, the object of wonder is a discrepancy between theoretical predictions and experimental findings, where both elements make up a scientific image of the world. In philosophy, the object of wonder is a discrepancy between the scientific image of the world and our conceptual intuitions about how the world looks, which builds the manifest image of the world. The goal of asking a philosophical question is to overcome this discrepancy, acquiring what Sellars (2007) calls a 'synoptic vision' in which the scientific and manifested image of the world can be reconciled. The means of achieving this goal – and solving a philosophical problem – is a conceptual description of the problem that will allow it to be solved with the help of scientific methods.

As is usually the case with philosophical confessions of faith, the aforementioned credo neither calls for nor provides much in the way of justification. However, if it is at least partially true – and I believe it is – it helps us to see what the philosophical problem is with thinking in general and thinking with images in particular. What makes that last issue philosophically interesting is a deep discrepancy between the scientific image and the philosophical understanding of what thinking is and what thinking with images can and cannot be. Examples of this discrepancy include the problems we encounter when we try to reconcile two seemingly justified beliefs: that the way engineers, chemists, biologists, architects or artists use images is essential to their acts of thinking and that the traditional way we think about thought and thinking, put in terms of inference relations and the content of propositional attitudes, is non-imagistic in nature. In other words, we have strong reasons to believe that we think with images and that this belief cannot be reconciled with the way we think about thinking. I believe that any attempt to overcome this discrepancy demands reconsidering our philosophical views on the nature of thinking and the role of images in the thinking process.

The main research question of this book is this: What is thinking with images? The question is analogical to such questions as 'What is thinking with language?' It means that if we can ask whether we can think in or with language, then we can ask whether we can think in or with images.[1]

The question follows from a commonsensical observation: when we ask how many windows are in the flat, someone will probably form and inspect a mental image of the flat and count the windows. If an architect designs a house, then they design the house with the help of drawings. One may use a map if one tries to get from point *A* to point *B*.

The aforementioned examples are instantiations of what can be called 'thinking with images'. However, listing examples of imagistic thinking is a relatively easy task. The difficult task is to say what thinking with images is.

Imagistic thinking is understood here in three ways: as a faculty, an act and a mental state or an event. The faculty of imagistic thinking refers to the capacity to use images in thinking. The act of imagistic thinking is exercising this faculty. The mental state called 'imagistic thought' is a product of such an act.

These three understandings of the term 'thinking' are interconnected, for we can only study the faculty of thinking through its expressions in the acts of thinking. The acts of thinking are interpretable only through studying their products. Similarly, we can understand what thoughts are only if we understand what the acts and the faculty of thinking are. Thus, a full-fledged theory of imagistic thinking should provide a theory concerning the faculty, the act and the mental state of imagistic thinking.

Imagistic thinking is commonly contrasted with thinking with words (e.g. Slezak, 2002b; Zhao et al., 2020), for, as is commonly held, not everything that is depicted or represented in an image can be described in language-like and propositional form. However, that is only a negative description of the phenomenon.

The main difficulty stems from the fact that thinking with images is not a case of trying to solve a clear-cut problem where we need answers to some ready-made questions. It is the opposite: we need a basic reconsideration and reorganization of the issue. The fact that we think with images is intuitive but poorly understood. It is partly caused by the fact that the concept of thinking with images has always been formulated in various ways, often just in ordinary language. The problem with spelling out what the expression 'thinking with images' might mean is a crucial part of what thinking with images is. Yet it is easier to point out what kind of definition we do not want than to produce one that we would accept.

Let me give an example of what such a definition cannot look like. It may be tempting to hold that iconic representation can be ascribed to the format of 'showing' and the propositional ones the format of 'saying'. However, it is a dead end for two reasons. On the one hand, one can overvalue the dissimilarities and claim that thinking with images expresses the so-called iconic difference, that is, it is governed by a 'logic of showing' which follows different rules (if any) than propositional logic – a 'logic of saying' (e.g. Belting, 2001; Boehm, 2007; Mersch, 2003, 2011; Mitchell, 1994; Müller, 1997). On the other, one may overestimate the analogies and claim that there is only one format of thinking – a propositional one (e.g. Fodor, 1975, 1987; Pylyshyn, 2003a). All other formats are only epiphenomena of the propositional one.

Both approaches are unsatisfactory. The first one is doomed to vague metaphors and cannot spell out the similarities between propositions and icons. Images and propositions can indicate a subject and attribute a property to it. If I take a picture of a man committing a crime, the picture can express a proposition that can be put into the sentence 'this man has committed a crime' (e.g. Kulvicki, 2020; Novitz, 1977).

The second approach is doomed to reductionism and cannot spell out the differences between propositional and iconic representations. Unlike images, propositions are governed by the rules of logic. Propositions can be negated and inferred. Images cannot. Thus, it is essential not to exaggerate the similarities and not to underestimate the differences.

Notice that the question 'What is thinking with images?' is not to ask 'Do we think with images?' One of the assumptions of the book is that we do. The question is, what does that mean, and what follows from it,

Although the nature and role of (mental) images have been investigated in many particular branches of philosophy, for example, in logic (e.g. diagrammatic reasoning), philosophy of mind (e.g. the imagery debate) and aesthetics (e.g. theory of depiction), what we certainly lack is a general account of what images are and what role they play in thinking.

The general form of the main question follows from our overall attitude towards what the philosophical problem is to begin with. As Sellars famously noted (2007, 369), the aim of philosophy is 'to understand how things in the broadest possible sense of the term hang together in the broadest possible sense of the term'. Thus, although the nature and the role of images are studied in many specific branches of philosophy, the general aim of philosophy is to deliver as broad an understanding of the nature and function of images as possible.

That is more than just a bold declaration or an expression of faith. Without an answer to the general question, we would not be able to avoid falling into a circularity of the following kind: two intuitions form the basis of the very idea of narrowing down the question 'What is thinking with images' to particular questions, such as 'What is the epistemological role of images?', 'What is the nature of depiction?', 'What are mental images?' and so on. First, one may claim that we can indirectly study the general nature of imagistic representations by studying how we use images within different domains. Second, it might be said that studying particular kinds of imagistic representations allows us to put the question of the role of images in thinking in a more precise manner. However, the claim that we can infer the general features of imagistic thoughts from studying their particular forms is based on a presupposition that we can distinguish certain forms of representation as instantiations of some representational genera. This possibility, however, requires that we possess at least operational knowledge of what imagistic thoughts are. Still, it is logically possible that such particular forms of imagistic thoughts instantiate heterogeneous representational genera. The last hypothesis, however, would be *undecidable* if we did not already have knowledge of what the nature of imagistic thoughts was in the first place. In other words, the point is not that we lack an answer, for example, to the question of the connection between mathematical diagrams, scientific graphs, mental images and pictures on a wall, or why we call them images. The point is that there cannot be an answer based solely on an analysis of particular phenomena.

The general approach to the question 'What is thinking with images?' seems to be rarely taken, with the notable exceptions of the works of Elgin (2017), van Fraassen (2008), Goodman (1976), Kulvicki (2014, 2020), Lopes (1996, 2003), Nanay (2014), Peirce (passim) and Price (1953), to name a few. Such general and synthetic approaches have always been exposed – most often fairly – to charges of oversimplification, ambiguity, misuse of metaphors, vague intuitions and analogies or simple ignorance. Such objections, which experts in narrow fields of expertise often formulate, often successfully undermine these enterprises. However, adopting a general approach is essential from the point of view of the development of science. It provides theoretical space for new hypotheses and inspires new research directions. Not to mention that it appears that we are currently stuck with no satisfactory theory of depiction or a universally accepted theory of the role of images in reasoning. We do not know what thinking is or how it relates to (mental) images. Thus, it seems reasonable to believe that what we lack is a synthesis. Some phenomena are visible only from a synoptic point of view. And this book is an attempt to do just that: to look at images and thinking from above.

There are two contrasting and firm intuitions standing behind the main question. First of all, there are instantiations of some mental processes, based, loosely speaking, on using images that we would rightly like to call 'thoughtful' but that do not fit into our general view of the nature of thinking. They do not have a propositional or language-like form, a form expressed with propositions or sentences.

For instance, we would certainly like to interpret architects' drawings as part of their thought process. And similarly, we may want to interpret how engineers, biologists and chemists use diagrams as an essential part of their scientific practices (e.g. Rheingold, 1991; Earnshaw and Watson, 1993; Nersessian, 2008; Tufte, 1997). Scientists think with diagrams (e.g. Ferguson, 1992; Gooding, 1990, 2010; Mößner, 2018; Sheredos et al., 2013; Shin, 2015, 2016; Sloman, 2002). Finally, we would like to say that the result of an artist's work, such as a painting or a piece of music, is a result of thoughtful action, even if such results are put exclusively in pictorial or auditory representational form (e.g. Carvalho, 2019). As Ryle (1979, 85–6) points out:

> Mozart's thinking results in something playable, not stateable. A symphony is not composed in English or German, it has no translation, there is no evidence for or against it. It is not grammatical or ungrammatical; neither in prose nor in verse. Cezanne may make mistakes, but he is not in error. There is no contradicting the chess player's carefully or carelessly thought-out move. A sonnet is not a report, a premise, or a conclusion; it can be a bad sonnet, but it cannot be a fallacious one.

All the mentioned examples of using images in science and art are hardly translatable into sentential forms, and it appears that they cannot be put into a logical form that would determine which step is necessarily next. Therefore, some acts of thought are hard to reconcile with our general view of what thinking is if we limit our understanding of thinking to propositional or language-based acts.

Second of all, there is a widespread assumption that images, particularly visualization, play an essential role in cognition and facilitate solving cognitive tasks. Most commonly, it is claimed that images are more informative and basic than sentential representations, epitomized by the maxim 'a picture is worth a thousand words', and that visualization fosters learning and reasoning processes. However, such generalizations about the benefits of images over good old-fashioned sentential representations beg the question of what is cognitively gained from them. What, for example, is the status of the commonly held claim that pictures convey more information than any formal representation or symbol? It is obviously false, for there are symbols, such as the symbols (1, . . ., ∞), that are infinitely more informative than any picture. It is not even the case that pictures necessarily convey more visual information. Some phenomena can be represented symbolically but not pictorially, such as the concept of an infinite set.

Moreover, it is far from clear what follows from the claim that visualizing helps in information acquisition (Salis and Frigg, 2020). For instance, how would visualizing the definition of an integral help us learn how to integrate a linear function? Visualization is neither necessary nor valuable in that context.

This tension is evident from the perspective of conflicting results in cognitive science in studies on the effects of visualization in reasoning. Some studies report that performance on some problem-solving tasks, such as those involving transitive inference, depends on the ease of imagining the premises of the reasoning (e.g. Clement and Falmagne, 1986; Shaver et al., 1975). Problem-solving task participants who build a physical model of a problem are more likely to solve it. That includes visualizing with the help of graphs and the help of gestures (e.g. Bocanegra et al., 2019; Goldin-Meadow, 2003; McNeill, 1992; Vallée-Tourangeau et al., 2016). However, numerous studies have failed to identify visualization as having any effect on reasoning. For example, Sternberg (1980) found no difference between the accuracy of solving difficult and easy-to-imagine problems. There is no correlation between the ability to visualize and scores on IQ tests and subjects' ability to reason. Richardson (1987) reported that individuals reasoning with concrete and easy-to-visualize problems were no better than individuals reasoning with abstract problems. According to some scholars (e.g. Byrne and Johnson-Laird, 1989; Knauff, 2013; Ragni and Knauff, 2013), images are irrelevant for reasoning. Human thinking is based not on abstract symbols or the manipulation of visual images but on spatial cognition, particularly on multimodal spatial layout models, which are more abstract than images and more concrete than symbols. It has even been demonstrated (e.g. Knauff, 2013; Knauff and Johnson-Laird, 2002) that visualization may, in fact, impede the process of thinking, which is known as the visual impedance effect. Knauff and Johnson-Laird (2002) show that visual imagery can slow down the reasoning process. In other words, engaging imagery skills can interfere with thinking.

The debate, however, is obviously unsolvable if we do not provide a clear-cut definition of what images and thinking with images are. For example, Knauff

(2013) rightly argues that mental imagery may actually hinder reasoning. However, the claim is only justified if mental images are understood as modality-specific, visual, conscious, picture-like representations that are supposed to play the same role as propositions. There is, however, no logical or nomological necessity for understanding them in this way. In fact, mental images are not necessary to be visual, conscious, picture-like representations or to play a proposition-like role.

We encounter the same problem when we assess the value of diagrammatic representations in the logic and philosophy of mathematics. Although it is rightly claimed that diagrams play a crucial role in mathematical practice, there is no consensus on the credibility of reasonings conducted with the help of imagistic means (e.g. Barwise and Etchemendy, 1996; Giaquinto, 2007, 2008, 2011; Giardino, 2017; Macbeth, 2012; Mancosu, 2005; Mumma, 2010; Nelsen, 1993). For example, it is hard to imagine Euclidean geometry without diagrams (e.g. Macbeth, 2010; Manders, 2008); simultaneously, diagrammatic representations have been suspected to be unreliable and heterogeneous (e.g. Brown, 1999; Giaquinto, 2011; Kulpa, 2009).

Moreover, the lack of a general understanding of the nature of images and their role in thinking may have a negative impact on our understanding of the role that imagistic representations play within cognitive systems. We risk making theoretical choices based on tacit knowledge and implicit assumptions about what images are and what role they can play in thinking without a clear account of what imagistic representations are. In other words, our theories of imagistic representations may very well be undercut by conceptual prejudices rooted in our vague intuitions. For example, it is argued (e.g. Pylyshyn, 2003b) that what hinders the imagery debate is the lack of agreement concerning what mental images can be and how we can explicate the phenomenon.

Consequently, we do not know what the empirical findings indicate. Without a comprehensive account of what 'mental image' means, we are condemned to be immersed in an abyss of vague metaphors and confusing concepts such as 'functional images' and 'quasi-pictures' that both are and are not images or a 'mind's eye' that is neither in the mind nor the eye. Additionally, it is argued (e.g. Berman, 2008; Reisberg et al., 2003) that the conflicting intuitions regarding the nature of mental images may be one of the causes of incompatible data in the imagery debate. It is plausible that preconceived theories about the nature of mental imagery influence what subjects say about their experiences, for theoretical views can and sometimes do influence introspection-based claims about imagery. Several studies (e.g. Intons-Peterson, 1983; Intons-Peterson and

White, 1981; Pylyshyn, 2002, 2007) demonstrate that subjects' performance in mental imagery tasks heavily depends on the subjects' expectations and preconceptions. Experimental results can be significantly distorted by cues regarding the experimenters' expectations, additional information and the subjects' intuitions (e.g. Hubbard, 1997).

The maxim 'a picture is worth a thousand words' does not seem to be informative regarding the relation between images and language either. It is commonly assumed that images are essential for using language.[2] For instance, in cognitive psychology, there is a well-established picture superiority effect which refers to the phenomenon in which images are more likely to be remembered than words (e.g. Defeyter et al., 2009).

However, without clarifying what we mean by images and words, it is hard to assess the relationship between them, interpret the available empirical findings and how these considerations might affect our understanding of language. Let me give an example of possible confusion this can lead to.

According to the research of Amit and colleagues (2017), the relation between words and images may be described as asymmetrical. We tend to generate visual mental images both in the case of visualization and inner speech, but we rarely think with words when we visualize something. The problem is how to interpret these findings.

On the one hand, one may say that it follows that manipulating iconic representations is a more fundamental operation than using language and that the meanings of the expressions of a given language are built upon the meanings of iconic representations. On the other, it may mean that imagery is essential in how we learn the language. For example, we can point at an image of a cat and express the proposition THAT IS A CAT, learning what the word 'cat' means. In this context, this asymmetrical relation between visual and verbal representations may mark the ontogenetic development of language. It may be a sign of how we acquire our first language, but it does not imply that it is an essential feature of the meanings of the expressions of the language in question.

Thus, the value and nature of imagistic representations cannot be assessed adequately based on our intuitions. Moreover, while there have been numerous empirical studies investigating different aspects of imagistic representations and significant theoretical progress on the topic since the end of the last century, few attempts have been made to integrate the findings into a coherent framework. As some researchers point out (e.g. Molitor et al., 1989; Scaife and Rogers, 1996; Winn, 1993), the idiosyncratic character of imagistic representations and their functions makes it difficult to generalize findings. For example, it is hard to apply the results of studies on the pictorial experience of mental images to motor mental images, and it is not clear how investigations of the nature of graphs might apply to pictures. It is difficult to make predictions about the cognitive value of mental images based on findings concerning the cognitive value of diagrammatic reasoning or to make predictions about pictures' cognitive function based on metaphors' cognitive function. What is needed is a systematic approach to interpreting the nature and merits of imagistic representations. Without such an approach, we have no principled way of making sense of the vast empirical literature on the cognitive function of (mental) images. Therefore, what is needed is both an answer to the general question 'What is thinking with images?' and a view that accounts for various uses of imagistic representations and the interplay between internal (mental) and external representations.

Image definition

Some remarks concerning basic terms are required. When I use the terms 'image' and 'imagistic representation', I interpret them as synonyms of the Peircean terms 'iconic sign' and 'iconic representation'.

According to Peirce (*CP* 1.545-567), representations can be divided into symbols, icons and indices. Symbols refer to their objects via conventions; icons share some qualities with their objects; indices are causally interconnected with the represented objects. This trichotomy is based on the form of the relation between the representation and its object. The distinction denotes different ways in which this relation can be founded, referred to by Peirce as the 'ground' of the sign (Short, 2007). We can speak of symbols, icons and indices, depending on the identified ground of the relation – an arbitrary convention, a similarity or a causal interconnection.

However, a decent interpretation of Peirce's theory of icons shows that it is disputable whether Peirce holds that the concept of resemblance is sufficient to explain how icons represent (Ambrosio, 2014; Chevalier, 2015; Hookway, 2000, 2007; Pietarinen, 2006, 2014; Stjernfelt, 2000). For the reasons mentioned later, it is better to understand the notion of 'image' as a representation that bears a natural and direct relation to the represented object (e.g. Burge, 2018; Giardino and Greenberg, 2015). This description is motivated by the observation that there is, without specifying it now, a natural relation between images and depicted objects which is believed to be based on exemplifying certain features of said objects. In comparison to the language, it is not accidental that a dog's picture is like a dog, where it is largely irrelevant whether we call a dog 'dog' or 'Hund'. The words 'dog' and 'Hund' are not dog-like; the image of a dog is. In comparison to natural signs, such as animal tracks in the snow, images do not have to be causally mutually connected with the represented object. An image of Santa Claus does not have to be caused by Santa Claus, while animal tracks have to be caused by animals. If there were no animals, there would be no animal tracks. We cannot say the same about Santa Claus.

Importantly, we cannot determine the sign's ground outside of the context of how the sign is taken. The letters p and b can be considered icons if we pick out the spatial similarities between them. They can be considered different symbols if we isolate the conventional relation by which they represent. Thus, depending on the interpretation of the relation between representations and represented objects, we can take these letters as instantiations of icons or symbols. Consider the diagram of a triangle $\triangle$. Depending on the way it is interpreted, it can be an icon; a symbol, for example, $\triangle ABC$; or an index, for example, a symptom of the mental illness manifested by compulsive drawing of triangles.

Following the Peircean tradition, I do not restrict the definition of images to visual representations. One can talk about tactile (e.g. Lopes, 1997; Kulvicki, 2006b; Yoo et al., 2003), haptic (e.g. Klatzky et al., 1991), olfactory (e.g. Bensafi et al., 2003, 2005; Djordjevic et al., 2004, 2005; Gilbert et al., 1998), gustatory (e.g. Croijmans et al., 2019; Kobayashi et al., 2004) and auditory images (e.g. Halpern et al., 2004; Hubbard, 2010; Jakubowski, 2020; Zatorre et al., 2010). They are just as common and just as psychologically important as visual ones (e.g. Sebeok, 1979; Newton, 1982). When addressing questions like 'How do you clean a window?' or 'How would you feel if you won a Nobel Prize?', one can form a motor (e.g. Guillot, 2020) and an emotional image (Blackwell, 2020; Nail, 2019; Thagard, 2005). We can use mental images to think about temporal relations (e.g. Viera and Nanay, 2020). Maps are imagistic representations of spatial relations. And diagrams are images spatially representing non-spatial relations. In a broad sense, even language is describable in pictorial terms. For instance, visual metaphors or graphic descriptions may be interpreted as literary devices that form a literary image or contribute to a literary image (e.g. Collins, 1991; Scarry, 1999; Troscianko, 2013).

Such a description is far from clear, and it comes as no surprise that there is no accepted definition of an image among psychologists and philosophers. That being said, an explication of the concept of an image is one of the aims of this book.

We can begin by introducing two constraints to help us produce a positive account. Based on this broad understanding of the term 'image', we can see, on the one hand, that any explanation of the imagistic genera should go beyond the experiential properties of images; on the other, it should be able to capture the role of the phenomenological properties of their content.

According to experiential accounts of depiction (Briscoe, 2016; Budd, 1995; Gombrich, 1960; Hopkins, 1998; Newall, 2011; Peacocke, 1987; Podro, 1998; Voltolini, 2015; Wollheim, 1987, 1998), perceiving objects displayed with images elicit a perceptual experience phenomenologically similar to seeing the represented objects in the flesh. However, experiential accounts are too restrictive to capture the concept of an image.

Let me illustrate it with Gregory's (2013) theory of distinctively sensory content. According to Gregory, the distinctively sensory content of images is subjectively informative. It identifies *what it is like* to be in a certain sensory state. It is perspectival; it specifies a perspective from which the sensory content of an image can be mapped onto the sensory content of an appearance.

The advantage of Gregory's definition is that it is broad enough to capture non-visual modalities. However, the definition of the content of images based on subjective informativeness and perspectivity is too narrow. First, these conditions seem to be insufficient. Suppose that a black canvas causes the common illusory experience of seeing Marilyn Monroe (Schier, 1986, 197). Still, it does not depict her. Second, we can speak of unconscious images deprived of phenomenological content. Traditionally, we define mental imagery in experiential terms (e.g. Finke, 1989; Thomas, 2003, 2009), but it is unnecessary to do so. It is logically and empirically possible to have a mental image we are not conscious of. It is logically possible since if there is no contradiction in the concept of soundless speech, then there is no contradiction in the concept of an unconscious image. It is empirically possible as there are subjects who are believed to have no conscious mental imagery experience and are still capable of performing mental imagery tasks (e.g. Zeman et al., 2010).

Moreover, if imagistic thoughts were only conscious experiences, we would be unable to capture unconscious mental events. Given that there are unconscious thoughts, it is implausible that all imagistic thoughts have to be experienced. And even if that were the case, the burden of proof would be on the proponents of the view that images are always conscious.[3]

With that in mind, any theory of thinking with images must assume that it is *possible* that there are unconscious images. Therefore, experiential properties of imagistic content cannot be a criterion for defining images.

However, abstracting away from the phenomenology of images is a dead end too. For instance, we can think of images in terms of the structural features of the format of representation (Dretske, 1981; Goodman, 1976; Haugeland, 1998; Kulvicki, 2006a) or in functional terms as props in so-called games of make-believe (Walton, 1990). Yet any theory of depiction should be able to explain the role of the experiential properties of images. If one were unable to do so, it would be considered insufficient.

Following the terminological practice of the theory of depiction (e.g. Kulvicki, 2014), I distinguish the term 'image' from 'picture'. 'Image' is a general term referring to all iconic representations. Pictures are a proper subset of images. When I use the term 'picture', I refer to external representations instantiated by photographs, paintings and drawings. Images comprise external (pictures) and internal representations (mental images) that are postulated to have iconic properties. The difference between internal and external representations lies in the nature of the vehicle of representation: either the representation is constructed in the mind, or it is based on an external and public medium, such as paper (e.g. Shin, 2002; Stenning and Yule, 1997).[4]

Imagistic thinking

The main problem with describing what thinking with images is stems from the fact that the general way in which images are described, including visual, auditory, olfactory, literary images, and so on, makes it difficult to read off the properties of imagistic thinking directly from the properties of images. In contrast, we can read off the properties of thoughts from the logical properties of language.

It is easier, however, to say what thinking with images is not. Imagistic thinking is contrasted here with propositional and language-like thinking. The term 'propositional thinking' refers to how we think about thinking and how we interpret the nature of thoughts. Thoughts are interpreted as contents of propositions. The thought that snow is white is the content of the proposition SNOW IS WHITE. Thinking is, first, the way we stand in relation to certain propositions, such as SNOW IS WHITE, which is most often interpreted as holding a propositional attitude, such as 'I believe that snow is white' or 'I doubt that snow is white' (e.g. Bermúdez, 2003; Peacocke, 1986),[5] and, second, as the way thoughts stand in relation to each other, which helps us to explain the nature of logical inferences between thoughts (e.g. Braine and O'Brien, 1998).

Propositional thinking is closely bound to language-like thinking. It is believed (e.g. Bermúdez, 2003; Devitt, 2006; Frege, 1984; Rey, 1995; Sellars, 1969) that for any propositional attitude, there should potentially, yet not necessarily actually, exist a sentence that makes the content of the attitude expressible. It means that propositional thoughts have linguistically expressible contents. It follows from the observation that thoughts are not directly interpretable. A thought has to be represented in an interpretable medium to grasp it. Language is the only medium that can mirror the logical structure of thought and express the richness of thought's content. Therefore, propositional thinking requires language.

In contrast, imagistic thinking is non-propositional, which means that it is not a matter of standing in some relation to a proposition and has content that is expressible in images. Imagistic thinking is irreducible to propositional thinking, just as it is believed (e.g. Ben-Yami, 1997; Crane, 2001; Grzankowski, 2013; Merricks, 2009; Montague, 2007) non-propositional states are irreducible to propositional ones. Thus, thinking with images is understood here as a kind of thinking that is non-propositional and expressible in images.

> *Thinking with Images* (*Imagistic Thinking*): the kind of non-propositional thinking, the content of which is expressible in images.

Paradigmatic examples of thinking with images are the way musicians use auditory images to think about music; the way painters use pictures to express thoughts about perceptual qualities; the way architects employ drawings to reason about spatial relations; the way scientists and engineers use diagrams and sketches in scientific reasoning; and the way people employ imagery skills.

According to the imagistic theory of thought, some thought processes are necessarily based on rule-governed mental or extramental image sequences. Images are the building blocks of some thoughts, and they determine the content of those thoughts.

A measurement-theoretic account of thinking with images

I will argue that the imagistic theory of thinking can be defended within the framework of measurement theory. The basic idea is that we can apply measurement-theoretic concepts to the analysis of imagistic thinking in the same way as we apply them to study propositional attitudes (see Dennett, 1987, 1991b; Dresner, 2010; Marcus, 1990; Matthews, 2007, 2011).

In a nutshell, I claim that thinking is an operation that locates an item in some logical or physical space, that is, an operation that enables us to localize and orient the item in relation to some parameters, such as volume, pressure, temperature, colour and shape, with the help of which we can recognize the item. Thinking is a manipulation of those parameters, or 'turning the knobs', as Dennett (2013) puts it. It is, for example, a matter of asking about what would change if some parameters were different, localizing the object of inquiry in a new measurement-theoretic set-up.

At the same time, thinking is not just any kind of operation. In contrast to dreaming and hallucinating, it is an operation that can be correct or incorrect (see Sellars, 1949). Consequently, imagistic thinking is a rule-governed operation. An analysis of imagistic thinking is an analysis of these rules.

According to the measurement-theoretic account of images, images can be interpreted in terms of being measurement devices, comparable to rulers and balances. Images exemplify constructions by means we localize objects. Consequently, I argue thinking with images is a rule-governed manipulation of construction rules revealing how something can be perceived in a measurement set-up, for instance, thinking about how an object would look if a parameter, such as height or length, were different. In short, thinking with images is a skill in using construction rules, comparable to using rulers. Thinking with images reveals the ways the world can be perceived and measured.

This sketch obviously requires further elaboration. It points us in the direction we are heading rather than presenting the final picture. The rest of the book is devoted to explicating just that.

However, one point raises no doubts. Studying the nature of images can teach us something important about thinking, in the same way as studying the nature of language reveals part of the nature of thinking. Thus, it is plausible that investigating the nature of images may play the same role in the philosophy of mind as investigating the nature of language. Suppose it is true, as it is claimed by most analytical philosophers from Frege to Dummett, that understanding the nature of language can bring us closer to understanding the nature of thoughts. Then, plausibly, the same can be said about understanding images.

The only difference is that although there is a well-established tradition of studying thought processes through studying the nature of language, similar analyses in the case of images are relatively underdeveloped. In other words, we know much more about language than about images and, consequently, about thinking with language than about thinking with images. This book takes a step towards changing that.

Content overview

Dennett has once said that philosophers are better at asking questions than answering them. I am not convinced that this is entirely true. However, I am certain that a clear formulation of a problem is a crucial part of a philosopher's toolkit.

Following this line, the book consists of two parts. In the first part (Chapters 1–3), I explain the nature of the problem of thinking with images. In the first chapter, I present the problems related to the idea of thinking with images. I argue that explaining the nature of thinking with images requires proving two theses. According to the Irreducibility Thesis, images are necessary for (some) thinking processes. This involves showing that images can be bearers of thoughts and express a non-propositional kind of content and knowledge. According to the Translatability Thesis, some imagistic thoughts and contents can be expressed by propositions. I hold that both theses can be explained and reconciled by adopting the measurement-theoretic account of images, according to which images are measurement devices, just like rulers and balances.

In the second chapter, I show how thinking about the nature of thoughts and images can lead to conceptual problems of thinking with images. Images cannot provide any theory of knowledge and content. They cannot explain the systematic and decomposable nature of thoughts. I introduce the so-called Received View, which is a set of beliefs and intuitions regarding the nature of thinking and which can be framed in a phrase that thinking is propositional and action-based. I demonstrate that the Received View is incompatible with the so-called Traditional View regarding the nature of images, which holds that images resemble represented objects.

In the third chapter, I explore the kind of answers we can expect by asking questions about the nature of thinking with images. I present two strategies for answering these questions. According to the neo-Lockean approaches, imagistic thoughts are abstract entities that mediate between perceptual and discursive representations. According to Wittgenstein–Ryle's line of argument, the question about the imagistic vehicle we think with is senseless. I show that both strategies fail. I introduce the so-called operational approach, according to which we can think of imagistic thoughts as the kinds of operations that require images to be expressed. The question about the nature of imagistic thoughts is the question of the nature of these operations.

In the second part (Chapters 4–7), I present a positive proposal for addressing the problem of thinking with images. I do this by analysing two case studies. In the fourth chapter, I study the content of knot diagrams. To explain how they represent, I introduce the concept of construction, which refers to the procedures of arriving at a target by determining the parameters of some logical or physical space. I show that the concept of construction is crucial for our understanding of how images represent.

In the fifth chapter, I analyse the black hole picture. In order to explain its semantics, I introduce the concept of recognition-based identification. I hold that recognition is distinct from having beliefs. It is a kind of reference that is also different from demonstratives and descriptions. Recognition consists in identifying construction invariants.

In the sixth chapter, based on the concepts of construction and recognition, I present the so-called two-dimensional (2-D) model of iconic reference. According to this model, images denote their targets and exemplify the rules of construction that identify the referent. I show that the best way to explain the semantical properties of images is to think of them as measurement devices.

In the seventh chapter, I show how the measurement-theoretic account of images can address the challenges described in the second chapter. I argue that the 2-D model provides us with a theory of knowledge and content. It allows us to explain the systematic and decomposable nature of thoughts. Last but not least, I show the metaphysical consequences the measurement-theoretic account of images has on our thinking about the nature of the mind. I apply the 2-D model to explain the nature of the representational format and mental imagery.

This book is long. Although I highly recommend reading it in the order presented in the consecutive chapters, it can be read in at least two ways depending on the subject of interest. Those who are intrigued by the philosophy of mind can go to Chapters 2–3 and 7. Those interested in the theory of depiction can focus on Chapters 4–6, where I present a model of iconic reference. However, this book should not be taken as presenting a full-fledged theory of depiction. That would require giving a detailed context of a discussion, explicating the positions I am arguing against and reasons why my model is better than others. The reader will not find it here since it would make the book unreadable. My goal is more moderate. I take the expression that we think with images for granted and investigate its consequences for our conceptions of thinking. I believe that the best way to make this expression sensical is by accepting the measurement-theoretic model of images.

1

What is the problem of thinking with images?

Trying to answer the question 'What is thinking with images?', we have to address two philosophical problems: the problem of the alleged *instrumental* role of images and the problem of the *epiphenomenal* nature of images. The first problem can be explained as follows: thinking with (mental) images is one of the recognized forms of non-verbal thought in contemporary cognitive psychology (Gardner, 2004; Kosslyn et al., 2006; Paivio, 1963; Shepard and Metzler, 1971) and neuropsychology (Greenberg et al., 2005), is basic for scientific reasoning (Abrahamsen and Bechtel, 2015; Bechtel, 2017; de Regt, 2014; Gendler, 2004; Meynell, 2018, 2020; Swoyer, 1991; Thagard, 2012),[1] is common in engineering (Ferguson, 1992) and in mathematics (Giaquinto, 2007), is indispensable for art (Arnheim, 1969) and is essential for common practical reasoning such as 'If I do [image], then I shall be able to do [image]' (Gombrich, 1990).

However, a time-honoured view is that the utility of images is only instrumental and cognitively inferior. Knowledge, traditionally conceived, consists of true and justified propositions. Knowledge is most commonly expressible in language, for only language can express logical relations. Images have limited expressive power; they are unreliable and cannot be trusted for inferences. In sum, they do not fit into a propositional account of knowledge. Images can only illustrate ideas or be used for the purposes of communication. Images can have a facilitating role and some psychological impact. But that is all. They do not have any genuine epistemological value.

One of the profound challenges in contemporary epistemology is to prove that images provide an irreplaceable and autonomous form of knowledge. Such a demonstration must not be confused with the dominant view that visualization is an inferior and illustrative presentation of cognitive processes and arguments, which have, for their own part, a source that is entirely independent of such visualization (e.g. Botterill and Carruthers, 1999).

The second philosophical problem with thinking with images concerns the alleged epiphenomenal nature of imagistic thoughts. Both problems are interconnected. First, if we accept the claim that the cognitive role of images is instrumental, then it follows that images may be epiphenomenal. They can accompany our thoughts while remaining inessential to the thoughts themselves and are ultimately unable to affect the contents of thoughts.

Second, if we agree that imagistic thoughts have an epiphenomenal nature, then it follows that they can only have an instrumental cognitive role. They may be widely used in cognition, much like pencils are widely used in engineering, but they are unnecessary for cognition, just as pencils are unnecessary for engineering. Images may be useful in cognition, just as pencils may be useful, but they do not ground cognition. They may prompt some thoughts, but they cannot determine the content of thoughts.

Historically speaking, there is at least one reason to believe that images play a crucial role in thinking. The so-called image theory of meaning and cognition was prevalent in modern philosophy and early cognitive psychology. It was believed that images could give us an answer to the question of concept acquisition. It was also widely believed that images possess iconic content, which characterizes perception.

However, both twentieth-century analytic philosophy and psychology are based on the rejection of images. Frege opens his seminal paper *Thoughts* (1984) with the constatation that images cannot be vehicles of thought. The roots of modern psychology can be traced back to the so-called imageless thought controversy and the birth of behaviourism. In *Psychology as the Behaviorist Views It*, Watson's (1913) example of imageless thoughts is given as among the first in justifying the critique of Wundt's introspection-based school of psychology. Due to twentieth-century critiques (e.g. Bennett and Hacker, 2003; Pylyshyn, 2003a; Ryle, 1949; Wittgenstein, 1953), it is now held that images cannot be representationally basic, and the phenomenon of thinking with images has come to be regarded as epiphenomenal and peripheral in cognition. The critique stems from two partially independent sources.

For one thing, it has been argued that images cannot determine the content of thought and that the central processes of thought require a propositional representation system. Images are considered epiphenomena of some abstract processes of thinking, conducted, for example, in some language-like symbol system.

For another, the view that thought might consist of disembodied images and that the mind is a sort of a private motion-picture theatre has always seemed philosophically dubious (e.g. Dennett, 1981a, 1991a; Goodman, 1984, 1990). Mental images can be studied by science, but as such they do not belong to the scientific image of the world. They are a part of its manifest image (Dennett, 1981b, 2017). In consequence, as was explicitly stated by Candlish (2001, 108):

> Now, however, it has become so thoroughly absorbed that it has disappeared from philosophy's surface and an imagistic account of thinking such as is outlined in Russell's *Analysis of Mind* (Lecture X) or elaborated in H. H. Price's *Thinking and Experience* is usually no more felt to deserve critical attention than is, say, a geocentric account of the universe.

Let me underline this point. The imagistic theory of thought has been replaced with the propositional one not because a competing theory with a more explanatory power has been introduced. The propositional theory of thought is, in many aspects, no better than the imagistic one. Instead, the imagistic theory of thought appeared to be internally inconsistent, and it was accordingly rejected. As Sellars (1963, 15) puts it: 'all attempts to construe thoughts as complex patterns of images have failed, and, as we know, were bound to fail.'

However, regardless of the staunch rejection of the imagistic theory of thought, even its harshest critics agree that even if not all thought processes are conducted in images, some of them are or at least that some thought processes do involve (mental) images. For instance, it is believed that images are crucial to explaining the nature of animal thinking (e.g. Burge, 2010; Gauker, 2011; Mullarkey, 2011) and the nature of concept formation (e.g. Barsalou, 1999; Carey, 2009). Whatever may be meant by the claim that mental images do or do not exist, mental imagery is a real phenomenon and quite open to quantitative scientific investigation (e.g. Shepard, 1990). It is claimed (e.g. Block, 1983a; Gauker, 2011; Kaufmann, 1996; Rollins, 1989) that the debate on the possibility of thinking with images challenges the dominant view of the nature of thinking. However, we still lack a clear understanding of what we mean by the imagistic theory of thought (e.g. Abel, 2012; Bechtel, 2008).

The Irreducibility Thesis

The imagistic theory of thought can be explicated in terms of the *Irreducibility Thesis*, which holds that images are necessary in thinking. It consists of the conjunction of the *Metaphysical, Semantical* and *Epistemological Irreducibility Theses*:

The Irreducibility Thesis: images are necessary in thinking.

The Metaphysical Thesis: Images are bearers of thoughts;

The Semantical Thesis: Images possess a kind of content distinct from that of non-imagistic representations;

The Epistemological Thesis: Images are vehicles of a kind of knowledge that cannot be obtained via non-imagistic means.

The Irreducibility Thesis expresses the belief that imagistic representations are neither epiphenomenal nor peripheral in the process of thinking. It means, first, that the existence of some thoughts depends on the existence of imagistic representations expressing these thoughts. Second, images and imagistic thoughts possess iconic content that is not reducible to other forms of content, in particular to propositional content. There is uniquely imagistic non-propositional content. Third, images and imagistic thinking do not provide merely instrumental and inferior forms of knowledge. Some cognitive functions are available only in virtue of the use of imagistic representations. It is not only that we can learn something from images, but images are a source of knowledge that is not available via non-imagistic means. The aim of the book is to show that the Irreducibility Thesis is true.

To illustrate the Irreducibility Thesis, think about how people keep photographs of their kids in their wallets. For some reason, we keep photographs of our kids and not name cards or their extended descriptions, although kids can be described with the help of propositions such as HE OR SHE IS BLOND, and so on. We want to say that the photograph represents something more. The question is what that something is.

One remark concerning the scope of the Irreducibility Thesis is needed. I argue that there are instantiations of the irreplaceable role of images, but I do not claim that every use of images is of that kind. There are certainly merely illustrative and instrumental uses of images. However, the Irreducibility Thesis holds that there are uses of images in accord with the thesis. Thus, even if images can sometimes be replaced in thinking, that does not mean that they are always replaceable.

Two interpretations of the Irreducibility Thesis

There are two ways of describing the irreducible role of images. First, we can understand the role of images weakly in terms of the facilitating role of images.

That means that the basic role of images is to prompt some thoughts and facilitate intellectual processes. Images can be interpreted as a necessary condition of some thought processes, where necessity is understood in epistemological terms: images are necessary to *grasp* some thoughts and acquire information. In the same way, computers are necessary to conduct some complicated calculations.

Second, we can understand the role of images in strong terms, where images are interpreted as constitutive elements of cognitive systems. According to the Strong Interpretation, the existence of some thoughts is necessarily bound with the existence of images. Necessity is understood in metaphysical terms, which means that it is impossible to *have* some thoughts without an image. As an example, numbers are necessary for counting; without numbers there would be no elements we could count with.

We can distinguish between two interpretations of the imagistic theory of thought with respect to each of the above. According to the Weak Interpretation, some cognitive processes, such as learning, reasoning or problem-solving, are based on iconic representations. For example, it is held that the use of mental images is crucial to solving mental rotation tasks or that there is a well-established relationship between the ability to use diagrams and success in mathematical problem-solving tasks (e.g. Casati, 2018; Grandin, 2006; Hegarty and Kozhevnikov, 1999; Thagard, 2005; Tversky, 2011, 2015).

> *The Weak Interpretation of the Imagistic Theory of Thought* (*Weak Interpretation* for short): Images are epistemologically necessary for grasping the content of some thoughts.

That being said, the Weak Interpretation is insufficient to explicate the imagistic theory of thought. Let me demonstrate that using Larkin and Simon's theory of diagrammatic reasoning.[2]

According to Larkin and Simon (1987), diagrams can be distinguished from sentential representations by the way they display information. Diagrams provide a one-to-one mapping of information stored in a spatial form at the particular locus of a diagram, including information about relations with the adjacent loci. We can infer the features of the represented objects by inspection of the spatial features of the representational vehicle.

In comparison to sentential representations, diagrams are cognitively less-loaded. They represent more pieces of information at once. They help in solving deductive and abductive tasks (e.g. Coopmans, 2014; Kirsh, 2010; Kitcher and Varzi, 2000; Latour, 1990; Zhang, 1997), for the process of acquiring information is perceptually enhanced by the way the information is organized (e.g. Bauer and Johnson-Laird, 1993; Beilock and Goldin-Meadow, 2010; Bordwell, 2008; Giardino, 2014, 2016; Ware, 2000).

Consequently, diagrams are the source of a 'diagrammatic' form of inference (e.g. Barwise and Shimojiima, 1995; Gurr et al., 1998; Lemon et al., 1999; Lindsay, 1998; Shin, 1994, 2002; Stenning, 2002; Stenning and Lemon, 2001; Stjernfelt, 2007). While symbolic reasoning is based on the interpretation of abstract symbols, a diagrammatic form of inference is based on the fact that we see the conclusion outright. What would be an active inference from premises to a conclusion in a symbolic representation system comes along as a 'free ride' in diagrammatic systems (e.g. Larkin and Simon, 1987; Shimojima, 1996, 2015).

According to Larkin and Simon, the difference between diagrams and sentential representations is a matter of different notation. Diagrams use spatial representations to express logical relations. Sentential representations are based on symbols. Yet, in principle, both representational systems can express the same information; they are informatively equivalent.

The difference in notation is not trivial. Some notations are more efficient than others, and much of the progress in mathematics is based on discovering different notational systems. Additionally, if the difference between diagrams and sentential representations is a matter of notation, it is relatively easy to prove that diagrams may be more efficient than sentential representations. The difference in the efficiency of reasoning can be interpreted in terms of the number of computational operations that need to be performed to move from the initial state to the goal state. Larkin and Simon's proof of the computational efficiency of diagrams does exactly that.

Although Larkin and Simon's theory succeeds in describing the epistemological merits of images, it is insufficient to explicate the imagistic theory of thought. Note that Larkin and Simon do not claim that diagrams are irreplaceable in thinking. Diagrams are good instruments for thinking with. They may be functionally necessary for some inferences, in the same way as computers are functionally necessary to calculate large numbers. Yet the same information that is encoded in diagrams can be expressed with symbolic representations. Sentential and diagrammatic representations are equivalent in the information they provide. They are not computationally equivalent.

The main problem with the instrumental view Larkin and Simon provide is that it does not exclude that images are parasitic on other forms of representations. Images can accompany the act of thought, they can help us grasping the content of thought, but they are not part of the logical mechanisms of thought (Kozak, 2020). Images, in accordance with Larkin and Simon's theory of diagrammatic reasoning, can efficiently make the content explicit. Images can draw attention to unnoticed properties of the content. They can prompt different ways of perceiving objects. Yet, they do not determine the content of thoughts. They can make some thoughts more precise and point out some features of the content, but they cannot change the content. In the same way, a microscope is a valuable tool to make evident the content of a sample, but it does not determine the sample's content. A microscope does not create microbes, but it may be necessary to reveal them. Larkin and Simon's argument demonstrates that images can play a crucial role in grasping the content of some thoughts but does not prove that we think with images.

Let us turn to the Strong Interpretation. According to the Strong Interpretation, imagistic thoughts can be distinguished by the nature of the operation. The nature of imagistic thought is interpreted in terms of a family of rule-governed processes. To ask about the nature of the operation is to ask what the operation does. For example, if one is interested in the nature of a quadratic function, then one is interested in how one operates with some variables. If one understands what a quadratic function is, one understands what it does. In the same way, if we are interested in the nature of imagistic thoughts, we are interested in what they do. In a nutshell, some operations could not exist if there were no images, in the same way as some mathematical operations would not exist if there were no mathematical objects. The question is what these operations are.

Accordingly, we can define the Strong Interpretation of the imagistic theory of thought.

> *The Strong Interpretation of the Imagistic Theory of Thought* (*Strong Interpretation* for short): Images are metaphysically necessary for the existence of some thoughts.

The Strong Interpretation holds that images are irreplaceable in thinking. The irreplaceability of images means that some thoughts do not exist, if there are no images expressing them. In other words, some thinking processes can be identified with genuinely imagistic operations. The main task here will be to establish this irreplaceable role of images.

The Translatability Thesis

One of the main problems with the Irreducibility Thesis is that it has to be reconciled with the Translatability Thesis. According to the Translatability

Thesis, imagistic thoughts can be translated into propositional ones. It means that some imagistic content can be expressed with the help of propositions. The idea is that we can easily go from images to words and the other way round. Words can be used to describe images. Images can be used to make statements, warn or criticize (e.g. Eaton, 1980; Kjørup, 1974; Kulvicki, 2020). Any theory of imagistic thinking has to take it into account.

The Translatability Thesis consists of the conjunction of two claims. The Semantical Translatability Thesis expresses the intuition that any content of an image can be expressed with a potentially infinite sequence of propositions describing the represented object. For instance, every map can be translated into a set of propositions in the form of verbal coordinates expressing the location and relations between spatial objects. The content of a picture can be described verbally, for example, with the proposition that it represents a woman with blond hair. Sounds can be expressed in a propositional form represented with notes, such as sound A follows B and so on.

The Epistemological Translatability Thesis states that knowledge expressed with images, such as 'if I do so [image], then I will be able to do so [image]', is expressible in propositional form, such as 'if I move the pedals clockwise, I will be able to ride a bike'. Knowledge of what my aunt looks like expressed with the help of a memory image can be translated into a description that my aunt is blond and so on.

The Translatability Thesis: Some imagistic thoughts can be translated into propositional ones.

> *The Semantical Translatability Thesis*: Some imagistic information can be expressed by propositional representations.
>
> *The Epistemological Translatability Thesis*: Some imagistic knowledge can be expressed by propositional knowledge.

The reason for accepting the Translatability Thesis is that we want to maintain correctness conditions for imagistic thinking, which means that we can be wrong when thinking imagistically (e.g. Dilworth, 2008; Langland-Hassan, 2015, 2020). First, we want to maintain the idea that it is possible to make a mistake in describing the content of an image. For example, it would be a mistake to say that a photograph represents a blond girl if it represents a dark-haired man. If propositions and images were untranslatable, then it would be impossible to say that they fit each other, in the same way as it would be nonsensical to ask whether a random set of letters describes a girl.

Second, we want to maintain the idea that imagistic representation may correctly or incorrectly represent the world. If I depict a wanted thief as a blond girl and the thief turns out to be a dark-haired man, I can say that I was wrong because the proposition THE THIEF IS A DARK-HAIRED MAN is true. If imagistic and propositional thoughts were untranslatable, it would be difficult to say in what sense I could be wrong when I depict the thief as a blond girl.

Most theories of imagistic thoughts are in tension either with the Irreducibility Thesis or the Translatability Thesis. There is a tendency to swing between the two. On the one hand, imagistic thoughts are sometimes believed to be a kind of propositional form of representation (e.g. Pylyshyn, 2003a), which makes it unclear how to express the genuine value of imagistic representation. On the other, imagistic thoughts are believed to be non-propositional (e.g. Gauker, 2011; Mößner, 2018), but it is unclear how we can translate the content of imagistic thoughts into propositional content. Hybrid views are often formulated (e.g. Camp, 2007, 2015, 2018; Denis, 1991; Fodor, 1975; Kulvicki, 2020; Langland-Hassan, 2015), which state that a proper part of the content of imagistic representation is propositional. However, it is not easy to see what, according to hybrid views, the contribution of the iconic content of thoughts is to cognition and what functions it plays in the broader cognitive economy.

Let me illustrate the problem with hybrid views using Fodor's theory of mental images. Fodor (1975, 2008) holds that images have to be put under description to fix their meaning which then allows them to be implemented in the mental mechanism. A sentence in the language of thought (LOT) has to be attributed to imagistic content. In a nutshell, if we introduce a system of mental symbols, such as LOT, and assign these symbols to images, we can determine the meaning of images and incorporate images into the machinery of thought.

However, Fodor's hybrid strategy renders imagistic representations redundant. If a discursive representation determines the content of images, then in principle, imagistic representations have no genuine role in cognitive architecture and can be reduced to propositions. That is, if we assign each image a sentence in LOT and define thinking operations as operations on LOT's symbols, then images do nothing in the mind's machinery. Images can play a role in grasping the content of some thoughts, but they are not necessary for having any thoughts.

In this book, I argue that any attempt to defend a non-trivial imagistic theory of thought has to take into account both the Irreducibility Thesis and the Translatability Thesis. It means that imagistic thoughts have to be expressible in propositions and have to carry such information that propositions cannot express. Most of this book will be devoted to showing how the Irreducibility and Translatability Theses can be comprehensively explicated in the context of the imagistic theory of thought. I shall argue that the former can be defended based on a measurement-theoretic account of imagistic representations. In a nutshell, we can think about images in terms of measurement devices. Therefore, we can defend the genuine role of images in thinking without being committed to the view that thinking consists in forming propositional representations of imagistic content.

Summary

Thinking with images is the kind of non-propositional thinking the content of which is expressible in images. The problem of thinking with images is how to reconcile a set of contradictory claims: on the one hand, it is claimed that images play an essential part in cognition and, on the other, that they are instrumental and peripheral in thinking.

According to the imagistic theory of thought, the existence of images is necessary for the existence of some thoughts. The imagistic theory of thought may be explicated in terms of the Irreducibility Thesis, according to which images are bearers of thought, possess a distinct kind of content and provide knowledge that cannot be obtained via non-imagistic means. The main problem with the Irreducibility Thesis is that it has to be reconciled with the Translatability Thesis, according to which some imagistic thoughts can be translated into propositional ones.

2

What is thinking?

Although thinking is a familiar phenomenon, there is no full-fledged theory of what thinking is. In philosophy and cognitive science, the issue of thinking is rarely tackled directly. Even if some authors do try to tackle the problem of thinking (e.g. Ryle, 1979), they usually deal with the problem indirectly, coming out with telling descriptions of neighbouring matters, but rarely coming out with a clear description of thinking alone.

It seems that any theory of thought should meet three requirements. First, it should be able to deliver a theory of knowledge, for knowledge is a structure based on thoughts. Second, a theory of thought should be able to deliver a theory of content, for we want to know how to determine what we think of. Third, it should be able to deliver a metaphysical theory of what the bearers of thoughts are.

These three requirements challenge any attempt to formulate a compelling theory of imagistic thinking. The scepticism concerning the concept of imagistic thinking is based on the belief that images do not have epistemological, semantic and metaphysical features that we are willing to ascribe to thinking.

At the same time, a loose set of beliefs and intuitions about thinking builds the so-called Received View of what thoughts and thinking are. The Received View is not an actual theory. It is rather a research programme rooted in the Kantian tradition. It consists of the belief that thinking is propositional – it is based on logical structures made of true-evaluable components – and action-based (Bayne, 2013).[1] The strength of the Received View is that it seems to meet the requirements for the theory of thinking. Moreover, it is in line with our deeply rooted intuitions about the nature of thinking.

However, the Received View seems to contradict the imagistic theory of thought. Thus, any attempt to create such a theory requires going beyond the Received View, that is, broadening the notion of thinking so that it can include imagistic thoughts, meeting the requirements of the theory of thoughts simultaneously.

The Epistemological Challenge

Any theory of thinking should be able to deliver a theory of knowledge. For although thinking and knowledge are separate phenomena, they are conceptually bound. One of the main reasons we value thinking is that it is a knowledge-building operation. It means that the main role of thinking is broadening our knowledge. Similarly, knowledge is not a matter of coincidence; it is a product of thoughtful actions – reasoning, deduction, hypotheses and so on. For all these knowledge-building operations, thinking is indispensable.

One of the main reasons for the scepticism surrounding the concept of imagistic thinking is the problem of its relation to knowledge. According to the sceptical line of argument, an imagistic theory of thinking cannot deliver a theory of knowledge, for images are not truth-bearers and do not have a logical form.

> *The Epistemological Challenge*: An imagistic theory of thinking cannot provide a theory of knowledge.

The Epistemological Challenge does not hold that images cannot be a source of knowledge. They obviously can. They can prompt some knowledge; they can justify modal (e.g. Williamson, 2016) and factual knowledge (e.g. Dorsch, 2016; Kind, 2018). Images can be direct objects of research. Images can be efficient tools for thinking (e.g. Larkin and Simon, 1987; Sloman, 1978, 2002). They can present procedural steps of reasoning in a more comprehensive manner and can be part of argumentation processes. Thus, images can be a source of knowledge, mainly because the human mind cannot grasp all information discursively. However, knowledge does not consist of images. In the same fashion, perception is a source of knowledge; and yet it is not an element building the structure of knowledge. According to the Epistemological Challenge, images have an instrumental nature – they help us acquire knowledge but do not constitute knowledge.

It is held that knowledge consists in providing information that something is the case. It is truth-evaluable and consists of elements that have truth-conditions. To have truth-conditions means to be a truth-bearer. Propositions are believed to be primary truth-bearers, which means that their content can be evaluated according to whether they are true or false. Knowledge consists of truth-evaluable and justifiable elements, such as propositions and logically structured relations between these elements. Thus, the knowledge that something is the case is traditionally dubbed propositional knowledge.

A propositional theory of thinking holds that thinking is a relation to propositions building propositional attitudes. It gives us an answer to the question of what the relation between thinking and knowledge is. If thinking is a relation to a proposition, then it is self-explanatory that it can be part of propositional knowledge. In contrast, the relationship between imagistic thinking and propositional knowledge is debatable.

On the surface, images can be true too. Images can work as arguments in scientific and legal practice. For instance, a photograph of a robber can serve as evidence in court. There are also times when we want to say of an image 'how true!' or assert that one image is closer to the facts than another (e.g. Eaton, 1980; Perini, 2005, 2012; Solt, 1989; Zeimbekis, 2015).

These intuitions, however, are clearly inconclusive. For to function in an argumentation process, an object does not have to be veridical. For instance, a knife can serve as evidence in court, but that does not mean that the knife is true. The knife makes the argument true, but it is not true in itself. In other words, images can function as truth-makers, but they do not have to be truth-bearers (e.g. Goodwin, 2009; Heck, 2007).

Moreover, one can posit that images are not truth-evaluable and that they can still provide true descriptions. Based on John's portrait representing him as bald, we can describe him as bald. It does not mean, however, that images are true or false. Similarly, a knife at the crime scene can provide a true description of what has happened. Yet it does not mean that the knife is true. We often say that photographs do not lie, yet this concept of truth therein is not the same as in the case of propositions. Images can be used to form true descriptions, but that is not to say that images are true or false in the same sense as descriptions are.

Still, we need an argument for why images are not truth-evaluable and cannot deliver a theory of knowledge. We also need to know what kind of argument we do not want. For one thing, one can argue that iconic content can be accurate or inaccurate (Crane, 2009; Greenberg, 2018, 2021). Accuracy comes in degrees; truth does not. A picture of John can be more or less accurate but not true or false. A proposition cannot be more or less true. Truth and falsehood are all or nothing. This argument, however, is unsound, for we can easily apply the idea of coming in degrees to a set of complex propositions.

For another, one can argue that iconic content is informatively rich (Dretske, 1981; Kitcher and Varzi, 2000) and finely grained (Tye, 2005). It means, first, that iconic content conveys too much information to express it with a finite set of propositions and, second, that iconic content is detailed and determined. In contrast, propositional content is general and abstract. This argument, however, is unsound. There are propositions that are rich in content, such as 'π equals 3,14 . . .', and images that are informatively primitive, such as an image of a dot. Further, propositional content can be more fine-grained than iconic content. Pictures of aqua and cyan objects are often not detailed enough to see the difference between them – the propositions 'x is aqua' and 'y is cyan' are.

In the next three sections, I present two positive arguments for a non-propositional nature of images: Wittgenstein's argument from content indeterminacy and Frege-Davidson's argument from lack of logical form. Both arguments will be the basis of semantical and metaphysical objections as well. Therefore, they will be presented here in a detailed manner.

Wittgenstein's argument from content indeterminacy

Wittgenstein's scepticism regarding the truth-evaluability of images concerns mainly the indeterminacy of iconic content. The argument is illustrated by a well-known remark from *Philosophical Investigations* (1953, §139): 'I see a picture; it represents an old man walking up a steep path leaning on a stick.—How? Might it not have looked just the same if he had been sliding downhill in that position?'

Wittgenstein holds that the same image can have various interpretations, depending on how it is taken. The walking-man image certainly represents some colour and spatial relations. But does it represent the man walking up or sliding down? In the case of language, you can determine the cognitive meaning of a sentence based on the meaning of its constituents and syntactic rules. The content of images is indeterminable.

The indeterminacy argument can be easily extended. Images cannot distinguish intensional, self-referential or modal contexts. A walking-man image can mean either that I believe that the man is walking, that I doubt that he is walking or that I hope that he is walking. The same image can express the thought that a man is walking and that a bank-owner is walking. It expresses the thought THE MAN IS WALKING, as well as the self-referential thought I THINK OF THE THOUGHT THAT HE IS WALKING. It may mean that it is possible that he is walking or that it is necessary. Images cannot distinguish these contexts; language can.

Let me clarify the matter. Wittgenstein's argument is not that images are ambiguous. Language is ambiguous, too. We can never know what a sentence means in a communicative context. However, we can assign different interpretations of the sentence to different propositions, which are not ambiguous. Wittgenstein's argument is metaphysical, not epistemological.[2] He argues that the meaning of images is indeterminable because there is always a method of projection that results in a certain image (Wittgenstein, 1958). To determine which method of projection is relevant to interpreting the content of an image, one has to know what the image represents. However, that is what was supposed to be determined by the interpretation. Thus, there can be no correct interpretation of the content of images, for the concept of correctness does not apply to it.

In consequence, images are insufficient to determine their truth-conditions and therefore are not truth-bearers; for to hold that representation is true, we need a systematic way to isolate the representational content and state the fact the representation refers to. Tarski (1956) accomplished that for formal representational systems by distinguishing the sequence of atomic symbols of the system and providing the means to translate them into metalanguage. Granted that Wittgenstein's argument holds, the same cannot be done for imagistic representational systems. If there is no systematic representation–reference relationship, then the concept of truth is not applicable.

Moreover, as Wittgenstein underlines, any way to counter this argument referring to some internal relation fixing images' content, such as inner knowledge of the meaning, makes this content inevitably private and vulnerable to private-language argument. If I knew what an image means, thanks to some internal feeling of knowing the meaning, then such knowledge would be unverifiable. Kripke (1982, 51) rightly calls this strategy 'mysterious and desperate'.

How should one evaluate the indeterminacy argument? Fodor calls it 'entirely convincing' (1975, 180) and uses it to argue that images cannot constitute thoughts. According to Fodor, to fix the content of an image, one has to name or describe it. For instance, titling a picture 'the walking-up-man' determines that the picture represents a man as walking up.

Yet the scope of the indeterminacy argument is limited (e.g. Goldberg, 2017). Wittgenstein's point is not that images are not bearers of thoughts and propositions are. For one thing, he puts too much effort into elaborating a pictorial way of describing ideas – in his works, he uses approximately 1,300 different images to present philosophical problems (e.g. Nyíri, 2013; Richtmeyer, 2019). For another, if images were deprived of meaning, then it would be senseless to speak of different uses of the same picture of an old man. In other words, one has to distinguish between the picture and its uses, in the same way as we distinguish between a rule and an application of the rule (Egan, 2011; Roser, 1996).

Thus, the goal of the argument is not to argue that we cannot think with images but that the *role* we attribute to images is misconceived. In other words, it is not the case that images cannot be bearers of thoughts but that no mental structure can play the role we attribute to images.

Wittgenstein's argument is directed against Russell's theory of image-propositions (List, 1981; Russell, 1919, 1921), according to which propositions can have either word form or imagistic form. The meanings of images are primitive – we know them directly and without reflection – and determined by the resemblance to the represented object. Accordingly, images are true if and only if there is a function mapping the image onto reality. For instance, a memory image representing John as bald is true if and only if John is bald.

According to Wittgenstein, if images were to serve a propositional-like role and their meaning were known directly based on perceiving similarities, then their contents would be indeterminable and could not be veridical. In particular, the resemblance relation cannot be epistemically primitive, for in order to know what aspect of resemblance is relevant for representation, one must know what is represented (Goodman, 1972, 1976; Suárez, 2003; Wollheim, 1980, 1987, 1998).

Yet, Wittgenstein's argument shows that images cannot play the same role as propositions, for we cannot determine the meaning of images based solely on resemblance relations. Moreover, the relation between images and thoughts is not analogous to the relation between language and thoughts. Images cannot play the same role as sentences. However, it does not support the claim that images play no role in thinking.

On the surface, two argumentative strategies can help us avoid Wittgenstein's conclusion. Both of them, however, fail. First, it is tempting to address Wittgenstein's argument by holding that we never approach images with no fixed preferences. After all, images are part of the communication process (e.g. Abell, 2009; Bantinaki, 2007; Blumson, 2009, 2014; Frixione and Lombardii, 2015). Therefore, they are never fully indeterminate. Some interpretations are more likely than others based on communicative context or knowing the author's intentions. For instance, Abell (2009), drawing on Gricean theory of meaning, holds that relevant aspects of resemblance and the content of a picture can be determined by knowing the author's communicative intentions. In short, a picture resembles an object O in the relevant respect if an author intends the picture to resemble O in this respect to bring O to the viewer's mind and intends that this resemblance have this effect because the viewer recognizes this intention.

However, this strategy is a dead end. We can appeal to the communicative context in the case of pictures. This strategy fails in the case of mental images. Internal representations are not part of any communicative practice.

We cannot determine the content of images based on the author's intentions, either. It would be circular to appeal to the author's intentions to determine the content of mental image. To know our own intentions, we need to know the content of mental images, but to know the content of mental images, we would need to know the content of our intentions. It would not be a problem if we could easily distinguish between the content of mental images and pictures. Yet, that would not provide a unified account of imagistic content.

Moreover, the intention-based accounts of depiction cannot explain how pictures represent. X-ray pictures are correct or incorrect regardless of the intentions of the author. Automatically taken pictures have content but not authors. There are unfulfilled intentions too. If I intend to draw a horse, but due to a lack of skill, the drawing is more like a cow, we would not say that the drawing represents a horse only because I intended to depict one. Therefore, the communicative context and author's intentions cannot solve the problem of content indeterminacy.

Second, we can claim that although pictorial content is indeterminate, it is not indeterminate all the way down. At least part of the content is determined. We can claim that there is some primitive content that is fixed and constrains possible interpretations of pictures. Kulvicki (2006a, 2014, 2020), following Haugeland (1998), calls it bare bones content. Bare bones content is defined in terms of projective invariants, that is, these representational features that are common for different interpretations of a picture, such as certain patterns of colours or spatial relations.[3] Bare bones content goes beyond the context of interpretation and captures basic semantic information displayed in pictures. It can be compared to the concept of character in the philosophy of language (Kaplan, 1989). In contrast, fleshed-out content is content that we can describe as a picture of John or a map of London and depends on our ability to recognize kinds of objects by the features of bare bones content. It can be compared to Kaplan's concept of content. According to Kulvicki (2006a), although pictures admit of alternative interpretations of the fleshed-out content, all these interpretations have a common bare bones content that constrains possible interpretations of the picture. For example, a square-like picture can represent a square or a face of the cube seen from a certain angle. These different interpretations mark the fleshed-out content. Yet the square-like shape is invariant in different interpretational contexts. It forms the bare bones content and constraints possible interpretations, for example a square-like shape cannot represent a circle.

This strategy, however, won't do. Primarily, we must determine which features are relevant as bare bones content. Some features, such as being painted on a canvas, are projective invariants but do not constitute bare bones content. Thus, to know which features are relevant, we must determine the class of abstraction.

However, in line with Wittgenstein's argument, we need to know what the picture's fleshed-out content is to determine the class of abstraction. Consider two examples. When interpreting a topological map of the underground, we must abstract away from spatial distances. To do that, however, we need to know that it is a topological and not a topographic map, which means that we have to know what the map represents – either topological or topographical relations.

Next, in the case of a black-and-white picture, we abstract away from the patterns of colorus. We are justified in doing so only if we know what the picture represents – is it a black-and-white picture of a coloured object or a coloured picture of a black-and-white object, such as a penguin? Thus, bare bones content cannot constrain possible interpretations of fleshed-out content, for to know what the bare bones content is, we need to determine the fleshed-out content.

Second of all, even if we acknowledge that bare bones content can be independently determined, it does not limit possible interpretations. We need to know more than just how to distinguish between accurate but different interpretations of given content. We need a differentiating criterion between accurate and inaccurate representations too. Obviously, there are inaccurate pictures. The question is how to distinguish them from accurate ones. For example, the same map can accurately represent London but also inaccurately represent Berlin. After all, having a map does not mean that we have a good map. However, the bare bones content of an accurate map of London and an inaccurate map of Berlin is the same. Thus, the indeterminacy of content is unavoidable.

Frege-Davidson's argument from lack of logical form

Frege's argument against imagistic thinking is based on images' inability to form logical relations. Logical relations are applicable only to truth-bearers. Propositions can be true or false, and the truth is the central semantic concept of propositional logic. As Frege (1984, 351) puts it: 'Just as "beautiful" points the way for aesthetics and "good" for ethics, so do words like "true" for logic.'

For logical relations to hold, the elements of the relation have to possess a logical form, that is, a syntactically fixed structure, such as a set of logical constants and variables, with determined transformational rules that preserve the logical values of its components. If *A* implies *B*, then based on transformational rules, it is possible to transform the truth of the first into the truth of the second. Propositional logic shows how the truth and falsehood of complex propositions depend on the truth and falsehood of simple ones. Truth-functions operate on propositions that can be negated, disjointed or conjoined; they can imply one another or be equivalent. One of the main reasons for talking about propositions at all is that they explain how things can stand in these logical relations.

Images lack logical form. They can illustrate logical relations, just like Venn diagrams can illustrate relations between sets, and apples can illustrate calculations. Images can help us to grasp logical transitions, just like apples can help us to grasp calculations. Yet they do not have a logical structure, just like apples are not logical entities. There are no truth-preserving transformation rules for imagistic representation. There is no pictorial negation (Crane, 2009; Sainsbury, 2005), conjunction or disjunction (Heck, 2007); images cannot express implications, logical modalities, quantifications (Frege, 1984) and so on. For instance, there is no image of all people being bald or such that it is necessary that they are bald.

Language is essential here. Grammatical constructions indicate relations between concepts and point to contradictions, consequences and correspondence between pieces of information (Loewenstein and Gentner, 2005; Lupyan et al., 2007). Images do not provide such opportunities.

Thus, images do not do any logical work. The best explanation of these facts is that images are not truth-bearers the transformational rules can be applied to.

According to Frege, the logical structure of thoughts mirrors the logical structure of language. Thus, the only route to an analysis of the logical structure of thought goes through an analysis of language. Dummett (1993, 128) calls it 'the fundamental axiom of analytical philosophy'.

The implication of Frege's argument is that it draws a clear line between images and beliefs. Beliefs are interpretable in logical terms; images are not. Beliefs are 'inferentially promiscuous' (Stich, 1978), which means that they can figure as premises in inferential transitions. It helps us to explain why images behave differently than beliefs. Images are not committed to truth; beliefs are. I cannot believe that $2 > 3$ if I know that it is the case that $3 > 2$. In contrast, I know that it is the case that there can be no flying horses, and yet that is not a reason to refrain from depicting one. Deprived of logical form, images are independent of beliefs.

We find a similar argument in Davidson (1997, 2001). According to him, thinking differs from other mental states in that it can be rational. A general theory of thinking should explain how it is possible to think rationally. Rationality is a matter of standing in logical relations. Beliefs are logically linked to each other in the form of a 'web of belief'. They justify each other and can rationalize one's actions. The only way to express these rational relations is through language. That is the main reason why Davidson holds that language is necessary for thinking (Davidson, 1982).

In contrast, lack of logical form renders images a-rational; they are neither rational nor irrational – the concept of rationality simply does not apply to them. If we think of images in terms of resemblance, then there is no room for asking about reasons. There can be questions about reasons when stating that p is q, but there cannot be a question of reasons when a picture depicts p as q. You can only show that, not argue it (Mersch, 2011).

Relations between images are not logical; these relations are usually understood as a causal chain of associations. The image of a mother can evoke a memory image of a family home, but the link between these two images is not a matter of logical consequence. We can speak of images' temporal or causal sequence, but rationality is not based on temporal or causal links. It is a matter of following logical rules.

Therefore, images cannot be rational. The difficulty is not that all thoughts are rational. They are certainly not. Some thoughts are irrational. However, if the concept of rationality does not apply to images, imagistic thinking is neither rational nor irrational. Thus, images cannot be constituents of thoughts.

The pictorial fallacy

At first glance, it may seem that the Frege-Davidson argument can be refuted by pointing out straightforward counterexamples. For instance, if I want to negate that John has red hair, I can depict him with blond hair. If I depict a green and a red apple, I express an alternative of a green and red apple. If one places two pictures next to each other, much like in a comic book, then one can say that their content is conjoined or that one implies the other (e.g. Malinas, 1991; Westerhoff, 2005).

These examples are misleading. The role of logical form is to determine the truth-conditions of its elements. In the case of images, truth-conditions cannot be determined. Having a picture of John with blond hair may be a negation of him having red hair, or black hair and so on; for the content of not having red hair is not simply being blond but an infinite alternative of the form 'having blond hair, or having black hair, or having green hair, etc.'. No image can represent infinitely many properties.

By the same token, conjunction, disjunction and implication do not simply represent a sequence of its elements; they set up a logical link between their components. In the case of two pictures, there is no way to determine the nature of this link – whether it is a temporal sequence, causal link, spatial transformation or simply a set of two unrelated pictures. In all these cases, the pictorial form is the same.

Moreover, images lack the generality and precision required by logical operations. For instance, it is impossible to depict the difference between the claim that $\exists x P(x)$ and $\forall x P(x)$ (Hintikka, 1987).

However, there is a more sophisticated line of argument available. It is based on two points. First, opponents of the propositional view of iconic content, such as Crane, usually agree that we can negate pictorial content. For instance, we can point at a picture representing John as bald and claim 'it is not the case that John is bald'. According to Grzankowski (2015, 2018), negation is a good way of testing for propositional content. If something can be negated, then it is truth-evaluable. Second, when seeing a picture, we can describe the picture's content in terms of modal propositions. For instance, when pointing at a picture depicting John being bald, we can assert that it indicates that in some possible world, it is true that John is bald (e.g. Kulvicki, 2020; Malinas, 1991; Matthen, 2014). Consequently, if the content can be negated and if we can say of the picture's content that it represents logical modalities, then it has a logical form.

This argument is invalid. Yet, it is invalid in an instructive way. Logical operators apply to propositions. They do not apply to the way pictures look. Thus, the picture's content must be put into a propositional form to which we attribute a logical operator. This is clearly possible. As Kaplan (1968) observed, many of our beliefs are of a form such as 'the colour of her eyes is [_]', where the blank '[_]' is filled with a pictorial representation. Hence, images can figure in what we believe and can play a role in forming propositional contexts.

Yet, the fact that something can be put into a propositional context does not imply that it has propositional content. Pointing at a knife can be used to express the proposition that 'it is (not) possible that *it* is the weapon Hamlet killed Polonius with', where 'it' is a deictic word, the referent of which is determined by indication. It does not mean that the knife is propositional. The proposition has been expressed deictically by indicating the knife. Still, the propositional content characterizes the deictic use of the word 'it' and an ostensive act of indicating the knife – not the knife as such.

This context can be multiplied, as in the case of deferred reference (Nunberg, 1993; Quine, 1968). Deferred reference is the use of an expression to refer to an entity that is not denoted directly by this expression. For instance, I can point at Quine's photograph and say that '*that* is most probably the greatest philosopher of the twentieth century'. The word 'that' does not refer to the photograph but to Quine. The visual properties of the photograph are used to recognize Quine or, putting it differently, to transfer the reference from the photograph to Quine. Similarly, I can point at John's portrait and express the proposition that it is possible that he is bald. Yet, the propositional content characterizes the deictic use of the picture, not the picture itself. The picture is used here in a deferred context. We represent John *as* bald to transfer the property of baldness to John.

Properties of representation are used to transfer reference to the represented object. In our example, the first attribution – representing John as bald – concerns the properties of representation and the second – attributing baldness to John – the properties of the represented object. Properties of representation describe the way the object is represented. Properties of represented objects describe the object itself. Only when we attribute properties to the represented objects can we speak of truth-conditions. That is why I can depict John as bald without being committed to holding the belief that John is bald.

Thus, we have to distinguish between the properties of representation and the properties of the represented objects. Properties of representation single out the way something can be represented. Properties of represented objects single out the way an object is or can be. The inability to distinguish these two kinds of properties is only another instantiation of the phenomenological fallacy. Let us dub it the 'pictorial fallacy'.

Let me illustrate the pictorial fallacy with the help of a drawing of a flying horse. If I draw a flying horse, I can attribute the ability of flight to the depiction of the horse. Therefore, I can say that it is possible to represent horses as flying, which is true. Yet it is something different than saying that it is possible that horses can fly, which is necessarily false in the same way as saying that it is possible that water is not H_2O.[4]

The argument that images can express modal and negative propositions is an instantiation of the pictorial fallacy. Note that the force of this argument depends on how we understand the concept of 'expression'. On the one hand, we can say that numerals express numbers, which means that numerals are necessary means to represent numbers. On the other, we can say that a book cover expresses the content of a book, which means that the book cover is an auxiliary means to understand the content of the book. The properties of the book cover represent the book's content and are used to recognize the properties of the book's content. In other words, the book cover is an illustration of the book's content. However, the book does not need illustrations to have content.

Pictures can illustrate modal and negative propositions. They can represent something *as* possible and negated, but they do not represent that something can be or is not the case; propositions do. Thus, indicating the fact that pictures can express modal and negative propositions does not imply that pictures have propositional content and can stand in logical relations.

The Semantical Challenge

According to well-established philosophical tradition, thoughts are intentional; they can stand for things, properties and states of affairs. In other words, they have content. Any theory of thinking should be able to deliver a satisfactory theory of content.

According to the same tradition, images are intentional, for they can represent the world by resembling it. Let us call it the Traditional View on the nature of depiction. According to the Traditional View, a picture of a red square stands for a red square for it resembles a red square. The problem with such a formulation of the intentionality of pictures is that if it were true, then an imagistic theory of thinking could not provide a theory of the content of thoughts.

> *The Semantical Challenge*: An imagistic theory of thinking cannot provide a theory of the content of thoughts.

Frege-Davidson's argument from lack of logical form and Wittgenstein's argument from content indeterminacy form the backbone of the Semantical Challenge. Both arguments imply that images cannot determine the content of thoughts and express predicative functions.[5] It shows that if the standard views of intentionality and pictorial semantics are true, then the content of imagistic thoughts is parasitic upon the existence of some more fundamental systems of representation.

One of the distinguishing features of thoughts is being intentional in *de re* sense. *De re* intentionality is most often compared to *de dicto* intentionality. Formally speaking, this distinction concerns ranges of modal operators; either *a* is such that it is possible that it is *P* or that it is possible that *a* is *P*. Different interpretations of the following expression can illustrate it:

(1) John believes that someone is following him.

According to the *de dicto* interpretation, John believes someone is following him. According to the *de re* interpretation, there is someone that John believes is following him (Quine, 1956). The truth of the *de dicto* interpretation of (1) would give a psychiatrist reason to be interested in John. The truth of the *de re* interpretation would give the police reason to investigate the case, for, unlike *de dicto* attitudes, *de re* attitudes attribute properties directly to the object of the attitude.

The different interpretations of (1) illustrate that having a de dicto attitude is insufficient for having a *de re* attitude. In other words, the belief that a proposition containing a term referring to something is true is not sufficient for having a belief about the object to which the term refers. The transition from *de dicto* to *de re* intentionality requires a special cognitive relation to the object (see Kaplan, 1968).

How should one think about this cognitive relation? Russell's idea (1997) was that the intentionality of thoughts (in the *de re* interpretation) requires that one has to know what one is thinking of. Evans called it Russell's Principle and expressed it as follows: 'a subject cannot make a judgement about something unless he knows which object his judgement is about' (Evans, 1982, 89).

According to Russell's Principle, intentionality *de re* is connected to knowledge of the object one attributes properties to. In other words, one cannot have thoughts of an object unless one possesses knowledge about the object one thinks about. Evans describes this knowledge in terms of discriminating capacities and dubs it 'discriminating knowledge'. The idea is that if one ascribes the belief FIDO IS A DOG to a cognitive system *S*, then it should be possible to ascribe to *S* a capacity to recognize the object the property BEING A DOG can be attributed to. Only if we can discriminate the object of our belief, namely Fido, can we predicate of the object Fido that it is a dog, which can be expressed with the sentence 'Fido is a dog'.

It does *not* imply that thinking that Fido is a dog is equivalent to having knowledge that Fido is a dog or that the act of thinking is equivalent to the acts of knowledge and cognition – one can think whatever one wants. It does not imply that the thought cannot be wrong – Fido could be a cat. It does not mean that we cannot fail in the ascription of thoughts either. As Freud might say, it may seem to us that we are thinking of a dog while we are actually thinking of our mother.

However, possessing discriminating knowledge means that having the thought FIDO IS A DOG entails a capacity to recognize the object one is thinking about – something which is necessary to report that one is thinking of Fido and not about anything else. Without such knowledge, there would be no object to predicate of and therefore there would be no thought at all.

Evans does not provide an explication of the principle. Rather, he leaves us with an example of a case where this principle is not satisfied. Suppose that a subject sees two steel balls continuously rotating about some point. At a later time, he reminisces THAT SHINY BALL IS PRETTY, but he does not know that in the meantime, he experienced amnesia that caused him to forget about one of the balls. In such a case, Evans holds that the subject does not have discriminating knowledge of the object he is thinking of; therefore, he has no beliefs about it, for he does not know which object can be described as pretty. As Evans states (1982, 115), 'there is nothing else he can do which will show that his thought is really about one of the two balls (that ball) rather than the other.'

Evans's position is definitely very strong and vulnerable to counterexamples (e.g. Hawthorne and Manley, 2012; Rozemond, 1993). For instance, I can think of the girl I saw yesterday in the park, even though I cannot distinguish the girl from her twin sister I had seen two days ago, not knowing which of the twin sisters I am thinking of. Even if I forget that I saw one of them two days ago, the name 'the girl I saw yesterday' fixes the reference of my thought. The idea is that we can pick out the referent and determine the content of thought without being able to discern the sisters; whichever sister I am thinking of, I am only thinking of the one I saw yesterday and not the one I had seen two days ago. I can be wrong in connecting this thought to one of the twin sisters in particular, but that does not mean I have no thought.

What does this counterexample show? It shows either that Russell's Principle is invalid or that Evans's argument in favour of the principle is invalid. However, we want to keep Russell's Principle for many independent reasons. For one thing, to veridically apply a predicate to an object, we must be able to identify the object (e.g. Burge, 2010). For another, to infer the properties of an object, we must be able to establish that the inference relates to the same object (e.g. Gersel, 2017), in which case, something must have gone wrong with the steel ball example.

Note that the main idea of Russell's Principle is that the object we predicate of can be individuated. The steel ball image, says Evans, does not allow us to individuate the referent. That, however, says something important about the nature of an image and its relation to thought and not about Russell's Principle itself. Although we have the conceptual capacities to individuate the steel balls – for instance, with the concepts THIS BALL and NOT-THIS BALL – we may lack the skill to apply these concepts successfully. We may fail in concept-application, but that is not particularly unexpected. I may think that the girl I saw yesterday was Jane, but it was Mary – her twin sister. It may happen.

In contrast, images do not provide us with tools to individuate objects we predicate of and it is debatable whether they can express predication. First, as Wittgenstein's argument shows, the content of images cannot be independently determined. The problem is not that we cannot be wrong about the interpretation of the content of an imagistic thought. The difficulty is that if the content of the images is indeterminable, then there can be no correct interpretation of the content. Thus, images cannot satisfy Russell's Principle. Propositions can.

Second, as Frege-Davidson's argument indicates, images lack logical form. As a consequence, we face the so-called binding problem of iconic representations. It is analogous to the unity of propositions problem in the philosophy of language (e.g. Gaskin, 2008). In the philosophy of language, the question is how to distinguish a proposition, such as THIS SQUARE IS RED, from a list of names 'this square', 'is' and 'red'. The answer comes easier if we assume that propositions have a logical structure that puts the names together. In the case of an image of a red square, it is either an image of a square, where the colour is irrelevant, such as in the case of black-and-white photographs, or an image of redness, where the shape is irrelevant, such as in the case of colour samples or, lastly, an image of a red square. The problem is that without a logical structure, it cannot be determined whether the content is combined or not.

Consequently, images are not capable of expressing predication, for predication requires a logical apparatus they lack. In particular, it requires a separation between arguments and predications. Propositional representations can do that. In the sentence 'it is a red square', we can point out the part that represents the argument and the part that corresponds to the predicate. In an image of a red square, the same part of the picture corresponds to the argument and the predicate. Consequently, a proposition that a square is red can predicate the redness of the square, for the square takes the place of a named argument we predicate redness of. In contrast, an image of a red square is in line with the interpretation that the square is red and that the colour red is square-like.

Let me clarify this point. Images surely bind contents together. An image of a red square combines RED and SQUARE and differs semantically from an image of a red triangle or a green square. However, the Semantical Challenge shows that we cannot determine the content of imagistic thoughts.

What I am suggesting here is that the Traditional View makes them vulnerable to the Semantical Challenge. That may be, however, an argument for changing our traditional way of thinking about images and not against an imagistic theory of thought.

The Metaphysical Challenge

It is believed (e.g., Davidson, 1997; Laurence and Margolis, 2012; Solomon et al., 1999) that concepts are the building blocks of thoughts. Why is that so?

As Frege-Davidson's argument shows, one of the distinctive features of thoughts is that they are systematically structured. Entertaining a thought of one kind entails a capacity to entertain a thought of another kind. For instance, entertaining the thought that JOHN IS HAPPY and that MARY IS SAD is systematically connected with the cognitive ability to entertain the thought that MARY IS HAPPY and that JOHN IS SAD. Having the thought that JOHN IS HAPPY entails a capacity to think that SOMEONE IS HAPPY. In Evans's words (1982, 104), 'if a subject can be credited with the thought that a is F, then he must have the conceptual resources for entertaining the thought that a is G, for every property of being G of which he has a conception.'

The same rule applies to inferences (e.g. Fodor and Pylyshyn, 1988). If I think IT IS DARK AND COLD AND RAINING, I can infer that IT IS COLD AND RAINING; for from P & Q I can infer that P (or Q). By the same token, I have to be capable of inferring from IT IS COLD AND RAINING that IT IS RAINING. If I cannot do so, I do not know what inference is.

Thus, thoughts must be systematically co-related (e.g. Heck, 2000; Peacocke, 1992), which means that they are systematic in nature. Evans (1982) calls this requirement the Generality Constraint.

To meet this requirement, thoughts have to consist of recombinable constituents that can build more complex structures. It means that thoughts are compositional in nature. The compositionality of thoughts means, first, that the meaning of a complex thought is determined by the meaning of its constituents. The constituents of thoughts are parts of thoughts that are canonically distinguishable, for not every partition of thought makes sense. The idea is that canonically decomposed parts are syntactically and semantically meaningful units. For instance, the thought that JOHN LOVES MARY can be decomposed into JOHN LOVES and MARY, but not into JOHN . . . MARY (e.g. Fodor, 2008).

Second, the meaning of complex thoughts must come from the meaning of its canonically distinguishable parts together with the rules of composition, for not all combinations are allowed. The recombination of the parts must be meaningful. JOHN LOVES MARY can be recombined into MARY LOVES JOHN but not into JOHN MARY LOVES.[6] These rules are recursive. If I have a thought that JOHN LOVES HIS MOTHER, I have to be capable of having a thought that JOHN LOVES HIS MOTHER'S MOTHER and so on. Putting it together, it means that thoughts have a recursive syntax that combines canonically distinguishable parts according to the combinatorial rules (e.g. Pagin and Westerståhl, 2010).

Systematicity and compositionality allow us to explain how thoughts can be productive. The productivity of thoughts means that we can entertain a potentially infinite number of different thoughts. Moreover, we can understand an indeterminate number of thoughts we have never entertained before. It all makes sense if we assume that thoughts are made of systematically structured and combinable elements. These elements are concepts, and the complex conceptual structures are propositions (Peacocke, 1992).

Language seems to be systematic and compositional. It has syntax and distinguishable syntactic and semantic parts. Thus, one can hold that thoughts consist of language-like representations (e.g. Devitt, 2006). In contrast, images lack systematicity and compositionality; therefore, they do not have metaphysical properties we are willing to ascribe to thoughts.

> *The Metaphysical Challenge*: an imagistic theory of thinking cannot provide a theory accounting for systematicity and compositionality of thoughts.

Images are neither systematic nor compositional. As Fodor shows, they lack syntactic structure and canonical decomposition.[7] Images lack grammar capable of generating infinite sequences of sentences (Eco, 1995; Wollheim, 1993). Therefore, iconic representations do not meet the Generality Constraint.

Fodor's argument (2007, 2008) takes the form of the so-called Picture Principle. According to the Picture Principle, iconic representations can be distinguished topologically: although images have interpretable parts, they lack canonical decomposition. It means, loosely, that we can cut up an image however we like, and each image-part will represent a relevant part of the represented object. Thus, every part of the representation represents some part of the scene represented by the whole representation (e.g. Green and Quilty-Dunn, 2021; Quilty-Dunn, 2016, 2020; Sober, 1976). In contrast, discursive representations have canonical decompositions, which means that they cannot be cut into pieces however we like. Discursive representations have constituent parts. For instance, the content of the proposition SNOW IS WHITE can be decomposed into the parts SNOW and IS WHITE but not into SNOW . . . WHITE, which means that the expression 'snow . . . white' does not possess independent semantical value. Thus, although images can be decomposed, they cannot be canonically decomposed. However, if they can be composed and decomposed however one wants, then they lack syntactical structure.

Two clarifications are needed. First, according to the Metaphysical Challenge, images can still be useful in thinking. They can prompt some thoughts or facilitate information processing. However, they cannot be bearers of thoughts, for bearers of thoughts must meet the requirements of systematicity and compositionality (Pylyshyn, 2003a). Propositions can meet these requirements; therefore, thinking has a propositional structure. Thinking with images can at most be built upon operations carried out with propositions; therefore, images are epiphenomena of propositional representations.

Second, one might object that images can exhibit systematicity. For one thing, map-like representations seem to be systematic (e.g. Braddon-Mitchell and Jackson, 1996; Camp, 2007). For example, a part of a map representing that London is west of Berlin also represents that Berlin is east of London. For another, as Matthen (2005) notes in criticizing Evans's Generality Constraint, if one is able to imagine a blue circle and a red square, then one is able to imagine a red circle and a blue square. In other words, if a representational system can represent multiple features together, it can represent different configurations of these features.

These objections miss the mark. The systematicity of thoughts comes paired with compositionality; for thoughts to be systematic, we have to be able to distinguish between the meanings of the constituents and the meanings of the complex structures. The thought JOHN LOVES MARY is built out of the concepts JOHN, MARY and LOVE, which can be distinguished as separate semantical units. In the case of images, such a separation cannot be produced, for they lack canonical decomposition.

Let us recall Wittgenstein's indeterminacy argument. According to Wittgenstein, the content of images cannot be independently determined. The metaphysical underpinning of Wittgenstein's argument is the observation that imagistic information lacks the syntactic structure that binds together the represented feature and the vehicle of representation.

Let me illustrate this claim. An image of a red square can represent the concept of REDNESS or SQUARENESS, as well as the proposition SOME SQUARES ARE RED. In other words, if one has a mental image of a red square, does it express the singular concept RED SQUARE or the proposition SOME SQUARES ARE RED? Language-like representations have syntactical structure and can distinguish between concepts and propositional structures made of concepts. Images do not.

To sum up, images lack syntactic structure, and, therefore, they do not meet the requirements of the Generality Constraint. They are neither systematic nor compositional. Any theory of imagistic thinking has to address the problem of systematicity and compositionality of thoughts.

The Received View

The epistemological, semantical and metaphysical challenges are met by the so-called Received View, according to which thinking is propositional and action-based. Moreover, the Received View seems to have more explanatory power than competing theories.

The propositional theory of thought stems from the rationalist tradition in philosophy. It is a set of positions that have in common the belief that the feature defining human thinking is rationality and that thinking itself can be modelled according to truth-preserving rules of logic (Fodor, 1991). Thinking is a set of operations on propositions and a matter of holding certain relations to propositions expressed in propositional attitude reports. Accordingly, thoughts are manifestations of propositional attitudes. Propositionalism in twentieth-century philosophy and cognitive science is represented by, among others, Frege's theory of thought, Fodor's LOT, Newell and Simon's GPS and mental logic theories.

Frege's theory of thought best expresses propositional theory. According to Frege (1984), thinking is defined as a subject's relation to thoughts. Thoughts are most often interpreted as abstract objects we refer to by means of representations of the type 'that-p' expressed in propositional attitude reports such as 'I believe that p' or 'I hope that p'. Thoughts are also distinct from the sentences that express them. Different sentences can express the same thought, such as in the expressions 'snow is white' and 'Schnee ist weiß'. The constituents of thoughts are concepts; in the same way, words are constituents of sentences expressing thoughts. Thus, someone who thinks that Venus is the second planet from the Sun must possess the concepts of VENUS, PLANET, SUN and 2.

Propositional theories are most often contrasted with associationist theories of thought.[8] The common root of associationism is the belief that rationality is not the best way to describe relations between thoughts. Thoughts are related to each other because, in their causal history, certain facts have causally linked the mental states into pairs, ensuring that if one pair member is activated, the next one is also activated. For instance, the frequency with which a certain organism came into contact with events X and Y determines the frequency with which that organism will have related thoughts about X and Y.

Hume's theory of ideas illustrates this well. It is primarily a theory of how perception (impressions) determines strings of thoughts (ideas). According to Hume (1975), ideas are copies of impressions in the following fashion: if the impressions IM_1 and IM_2 are related in perception, then the corresponding ideas ID_1 and ID_2 are also related. We do not need to refer to any intermediary entities like implicit rules. The causal order of perception thus determines the order of thought.

Associationism has many advantages. It explains the psychological mechanism of thought acquisition well. It is metaphysically unencumbered and does not postulate any hidden logical mechanisms. It explains many mental phenomena too. It is used in learning theory (e.g. behaviourism), theories of reasoning (e.g. dual-process theories) and theories of thought implementation (e.g. connectionism). It is in line with nowadays theories of mind, such as Enactivism and Dynamic Systems theories. All these theories are independent but share an empiricist core.

Associative structures are usually contrasted with propositional structures, in which individual elements are logically correlated. Propositional structures are not just a causal sequence of associations. Associative structures express only causal relations between representations, for example, the associative structure GREEN-TREE says that we associate GREEN with TREES; it does not express the proposition that (SOME) TREES ARE GREEN. The thought SOME TREES ARE GREEN does not tell us that there is a causal relationship between the GREEN-TREE representation. Instead, it predicates of some trees that are green.

Accordingly, associative inferences are transitions between thoughts that do not follow from logical relations between elements of thoughts. In this sense, associative inferences are contrasted with logical inferences, such as those made use of in the computational-representational theory of thought, in which inferences are truth-value-preserving transitions between thoughts that are determined by their syntactic properties. Associative inferences are based not on syntactic properties of thoughts but on associative relations between the contents of thoughts. For example, we can associate the thought LONDON and IT IS RAINING because we once got wet in London. We cannot infer that it is raining in London based on the formal properties of LONDON and IT IS RAINING.

However, associationism does not appear to be a satisfactory account of thinking in general. First, it is doubtful whether associationism can explain the predicative nature of thinking. The thought TREES ARE GREEN predicates something of trees. It does not just causally connect thoughts about trees and colours. Associative structures cannot explain how thoughts can predicate something of the world.

Second, the same thoughts can appear in different intensional contexts. I can believe that-*p* and hope that-*p*. Associationism can explain it only if it interprets the thought 'that-*p*' as different across various intensional contexts, which is what we are trying to avoid.

Third, thoughts are compositional and systematic (e.g. Fodor and Pylyshyn, 1988). The associative structure of thought that binds together the list of representations TREES, ARE and GREEN is compositionally indistinguishable from the thought TREES ARE GREEN. However, these are two different structures. Moreover, understanding the thought JOHN LOVES MARY presupposes that we also understand the thought MARY LOVES JOHN. It is unclear how to explain this systematicity in a model in which, for example, past experiences have not linked Mary and John to a mutually loving relationship.

Fourth, some thoughts are coextensive. For instance, the thought that CICERO IS A ROMAN PHILOSOPHER is coextensive with the thought that MARCUS TULLIUS IS A ROMAN PHILOSOPHER. Associationism captures the difference between these thoughts. It is not able to explain what the identity of these thoughts consists in.

In contrast, propositionalism explains rationality and the possibility of modelling thinking according to logical rules. It explains the phenomenon of systematicity and the compositionality of thoughts. It gives us an elegant description of how thoughts can appear in different intensional contexts. It provides an answer to the question of coextensionality and the predicative nature of thoughts.

Yet the greatest strength of propositionalism is the weakness of the competing theory, for accepting the propositional theory of thought does not come without a cost. First, it forces us to make a number of metaphysically loaded assumptions, such as that humans are rational beings or that rationalism best describes the phenomenon of thinking. Consequently, we must explain the place of rationality and thought in the physical world. Frege's explanation leads to an acceptance of Platonism; Fodor's view leads to the endorsement of the idea of an innate language of thought. Second, propositionalism cannot capture the phenomenon of non-propositional thinking, such as imagistic thinking.

How does one find a place for imagistic thinking within this division? There are two dominant strategies. First, we can try to interpret images in propositional terms, trying to adjust images to the requirements set by propositional theory (e.g. Blumson, 2012; Camp, 2007). Second, we can bite the bullet and claim that imagistic thinking is more appropriate to associationist theories of thought (e.g. Quilty-Dunn and Mandelbaum, 2020).[9] However, both strategies are unsatisfactory. On the one hand, propositionalism cannot provide a theory of imagistic thinking, for images are non-propositional in nature. On the other, adopting an associationist theory of imagistic thinking is running away from the problem.

Granted, associationism is not a false theory. After all, thinking can take different forms. When I think that TREES ARE GREEN, I can infer that SOME TREES ARE GREEN, but I can also think of summer. However, one wants to choose propositionalism over associationism because the first can provide a theory of rational thinking while the second cannot.

We want to ensure that rationality may be expressible in imagistic thinking since images are crucial for science and can work as arguments in reasoning. Moreover, according to the Translatability Thesis, we should be able to transform images into propositional forms and model relations between images according to the rules of logic. We certainly do that. Operations on sets can be performed on both Venn diagrams and propositions. Any theory of imagistic thinking should be able to include inferential transitions between images. Therefore, it has to respond to the challenges set by propositionalism and not be left at the mercy of associationism.

What is the lesson? It seems that neither propositionalism nor associationism offers a theoretically respectable way of incorporating images into a coherent theory of thought. Thus thinking with images cannot be explicated within the propositionalism–associationism distinction. However, this problem is deeper. It stems from the impossibility of freeing oneself from the traps of metaphors that determine our thinking about thinking.

Metaphors of thinking

Although the distinction between propositional and associationist theories has dominated philosophical discussions, it is often forgotten that it is underpinned by a more basic understanding of the nature of thinking. It is expressed through intuitions and metaphors rather than sets of theories. For this reason, it is rarely directly tackled.

There are two dominant metaphors for what thinking is that are at the heart of different theories of thinking (Schooler et al., 1995). According to the moving-through-space metaphor, thinking involves moving through a logical or physical space. Thinking is something we do and for which we can determine certain methods and rules in the same way as we can determine pathways in space. We use the spatial metaphor when we use such phrases as 'searching one's mind', 'approaching the problem from a different angle' or 'changing the direction of thought'. We are also willing to answer the question 'What are you doing?' with 'I am thinking'. According to the spatial metaphor, thinking is a kind of activity at which we can be better or worse. We can be 'too tired to think' or we can be 'deep' or 'slow thinkers'.

According to the perceptual metaphor, thinking is an act that happens to us, often unconsciously and automatically. Thinking is a kind of seeing, not doing. We use the perceptual metaphor when we use such phrases as 'seeing a solution', 'gaining insight into the problem' or 'casting light on something'.

These two metaphors do not overlap with the distinction between propositional and associationist theories of thinking. Frege, computational theories and behaviourism adopt the spatial metaphor. The perceptual metaphor has been developed within theories of divine illumination, the Cartesian theory of clear and distinct ideas, Locke's theory of ideas and Gestalt theory (Figure 2.1).[10]

The moving-through-space metaphor lies at computationalism's heart (Fodor, 1975; Newell and Simon, 1972). Solving a problem can be seen as moving through a problem space from an initial state to the goal state. Movement through the

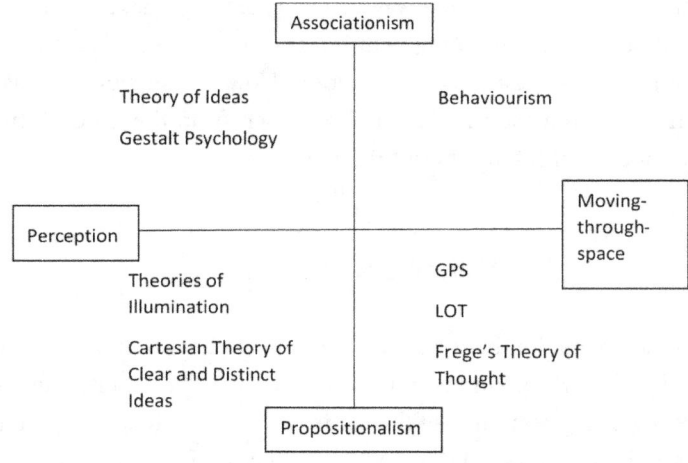

Figure 2.1 Theories of thinking. © Piotr Kozak.

problem space requires the use of operators. These are actions that fulfil certain subgoals. Moreover, holding propositional attitudes can be interpreted as occupying a place in logical and physical space and as rule-governed transitions from one state to another. For instance, the belief that grass is green can be considered a state in a logical space isomorphic with the corresponding physical exemplar of the sentence 'grass is green' at a particular location in the human brain. This sentence may be the basis for further computational processes comparable to movements through the state-space, such as the belief that my lawn is green, which, in turn, can be the basis for action, for example, to water the lawn.

The same moving-through-space metaphor forms an underpinning of the behaviourist theory of thinking. Behaviourists, such as Skinner (1957), held that mental states could be interpreted as abbreviations of certain actions. For instance, the feeling of sadness is an abbreviation of certain external behaviours, such as crying and facial expressions, and internal behaviours, such as neural activations, muscle tension and so on. In this vein, they assume that thinking is a 'silent speaking', manifesting in subtle behaviours such as laryngeal muscle movements.

However, the spatial metaphor does not seem to fit theories of imagistic thinking. Entertaining an imagistic thought of a sunny day in summer does not seem to be a movement in a logical space. Instead, it is a form of perception. In particular, the theory of thinking as silent speaking does not seem to apply to musical or visual thinking acts. When a painter thinks of a landscape, his thinking process takes the form of perceiving an image rather than speaking to himself.

The perceptual metaphor is storied. It has its roots in the Aristotelian theory of intellectual intuition. It has been used, among others, in medieval theories of divine illumination, the Cartesian clear and distinct ideas, Locke's theory of ideas and Gestalt Psychology.

However, the perceptual metaphor is at loggerhead with what we know about perception and thinking. On the one hand, philosophical theories cannot explicate the difference between perception and thinking (more on that in Chapter 3). On the other, even though it is promising, Gestalt Psychology is far from providing a comprehensive theory of thought.

Gestaltists analyse thinking in problem-solving contexts and emphasize insight's role in understanding a problem's structure. Within Gestalt Psychology, understanding is not an incremental and continuous process but is sudden and spontaneous, similar to an act of perception. However, Gestalt Psychology is fairly accused of being uninformative (e.g. Jäkel et al., 2016). It does not explain phenomena but describes them. It does not indicate the mechanisms governing the thinking process but describes the qualitative structure of the thinking process. It does not explain, for example, what a sudden process of insight consists in. Thus, it cannot be a candidate for a theory of imagistic thinking.

Where do we stand? On the one hand, the spatial metaphor is inconsistent with how we use images in thinking. On the other, the perceptual metaphor does not explain what a theory of imagistic thinking could be. However, it is forgotten that these two metaphors are not mutually exclusive. Although they are frequently discussed independently, they are complementary.[11] When moving through a physical space, we have to see where to go next. In problem-solving, movement from an initial state to the goal state requires recognition of the correct operators. Thus, these two metaphors can be readily combined. With respect to thinking, such a combination can be beneficial because it suggests that multiple processes may contribute to what we call thinking. It also suggests that numerous factors may contribute to an impasse in thinking. On the one hand, one can know how to move through a logical space but fail to see the goal. On the other, one can see where to go but not know how.

A more adequate metaphor for thinking would be that of measurement. Thinking is like applying measures, and thoughts are the products of those applications. The reasoning is to see what results from applying different measures.

The measurement metaphor contains the perceptual and spatial metaphor. Measurement is the result of a certain action. For instance, it may be the action of measuring the time between thunder and a flash of lightning. It is moving through logical space by determining spatial and temporal relations between measured objects. Yet, to conduct a measurement, one has to recognize what is to be measured – in this case, distance from the storm. With this metaphor in hand, we can consider alternative ways of building theories of imagistic thinking.

Summary

The Received View is a loose collection of beliefs and intuitions about the nature of thinking, according to which thinking is propositional and action-based. Thinking is a matter of operations carried out on propositional structures made of concepts.

The Received View gives a clear idea of what knowledge and the relation between knowledge and thinking are. It provides a theory of the content of thoughts. It is in line with metaphysical assumptions concerning the nature of the bearers of thought. Thoughts, as well as propositions, are compositional and systematic.

Imagistic representations do not fit into the Received View. They seem to be unable to provide a theory of knowledge and a theory of content. They seem to fail to explain the systematic and compositional nature of thoughts. Thus, if we accept the Received View, the concept of thinking with images is self-contradictory.

Any theory of imagistic thinking requires broadening the notion of thinking. At the same time, it should be able to address the epistemological, semantical and metaphysical challenges directed against such a theory.

3

What answers should we expect?

Bertrand Russell noted (1919, 11) that 'If you try to persuade an ordinary uneducated person that she cannot call up a visual picture of a friend sitting in a chair, but can only use words describing what such an occurrence would be like, she will conclude that you are mad'. However, the claim that we think with images has always been subject to suspicion, for our prescientific intuitions, which supposedly support imagistic theories of thought, have always been vague and fallible. That is a simple consequence of our lacking a clear understanding of what these intuitions are thought to support. The imagistic account of human thought is a source of misunderstandings and disputes mostly because it is not clear what we really mean when we say that we think with images. As Pylyshyn (2003b, 113) observed, despite claims that thoughts have a picture-like format having persisted for such a long time, the problem of stating clearly what it means for thought to be imagistic has rarely been explicitly tackled. Yet the real difficulty lies deeper: we do not even know what kind of answers we expect when we ask, 'What is thinking with images?' To know the content of a question is to know what counts as valid and invalid answers to the question. To know the content of the question 'What is thinking with images?' is to know what kind of answers we should expect. In other words, we should know what counts as an explanation of what thinking with images is.

We have to be careful not to trivialize the answer. We face the problem of triviality when we try to describe a phenomenon without explaining it. The problem of triviality stems from the intuitive character of imagistic theories of thought. At first glance, it seems like there is nothing wrong with imagistic thoughts and that they do not need any clarification: the claim that we can think with images sounds intuitive, and we have to accept it as a brute fact. This prescientific intuitiveness is the cause of our continued attachment to the imagistic theory of thought, but it is also a significant source of conceptual confusion.

This commonsensical intuition found its early scientific expression in Francis Galton's (1880) famous questionnaire *On Visualizing and Other Allied Activities*, which analyses, among other things, different styles of thinking. It would go on to receive further scientific attention in a number of studies on the nature of memory in the early 1960s (Baddeley and Lieberman, 1980; Baddeley, 1988; Paivio, 1963; Shah and Miyake, 1996) and research conducted in the chronometry paradigm (Shepard and Metzler, 1971), building what has been commonly called 'the pictorial approach' to the nature of mental representations (e.g. Kosslyn et al., 1994, 2006; Pearson et al., 2015).

Considerable worries arise when one tries to accommodate the trivial fact that images play an important role in cognitive practice with philosophical accounts of the nature of thinking. For from the fact that one commonly uses images, it does not follow that one thinks with images, just as the observation that we frequently use commas in writing says nothing about what language is.

Therefore, without any supplementation, the intuition that we think with images is either *trivial* or *false*. Either it says only that we commonly use (mental) images in cognition, which nobody disagrees with, or that from the frequent use of images in thinking we can directly infer the true nature of thinking, which is false. Without establishing what kind of contribution images make to thinking, nothing follows for the explanation of the latter. We need an argument as to why we consider using images an instantiation of thinking and not something that accompanies thinking.[1]

There are two dominant non-trivial strategies for answering the question 'What is thinking with images?', depending on how the notion of the bearer of thought is understood. According to the neo-Lockean approach, imagistic thoughts are theoretical objects that mediate between perceptual and discursive representations. According to Wittgenstein–Ryle's approach, the question is senseless, for there are no objects we think with.

In this section, I argue that both strategies fail to address the problem of what thinking with images is. I introduce an alternative strategy for answering this question that is put in terms of the operational approach. According to it, the imagistic bearer of thought can be understood in terms of the family of operations.

Imagistic thoughts as theoretical objects

The leading cause of the revival of imagistic theories of thought in the twenty-first century is a belief that a full-blooded theory of thought should explain how we acquire thoughts. According to the explanatory strategy represented by Locke[2] and neo-Lockean approaches, such as Barsalou's (1999), Gauker's (2011) and Prinz's (2002), images offer a plausible answer to the problem of thought acquisition. The basic idea is that we have no choice but to accept the imagistic theory of thought to solve empirical and philosophical problems concerning the nexus between perception and thoughts. According to this strategy, we posit the existence of some mental objects, such as imagistic thoughts and Lockean ideas, just as we hold in astrophysics that to solve problems with the accelerating expansion of the universe, we postulate the existence of dark energy. In both cases, we claim that images and dark energy are necessary for explaining certain psychological and physical phenomena. We postulate the existence of some physical and mental objects to make some phenomena explainable. The explanation of what imagistic thoughts and dark energy are would be similar: imagistic thoughts and dark energy are theoretical concepts that we adopt to explain certain phenomena. Thus, we need to know what problems images solve and how.

As Locke states in the opening chapters of his *Essay Concerning Human Understanding* (1975, 104–5):

> Let us suppose then the mind to be, as we say, white paper, void of all characters, without any ideas: How comes it to be furnished? [. . .] Whence has it all the materials of reason and knowledge? [. . .] To this I answer, in one word, from experience. [. . .] First, our Senses, conversant about particular sensible objects, do convey into the mind several distinct perceptions of things, according to those various ways wherein those objects do affect them: and thus we come by those ideas we have of yellow, white, heat, cold, soft, hard, bitter, sweet.

To understand the Lockean innateness-free strategy of argumentation, one must bear in mind that Locke's ontological argument is a consequence of an epistemological one. Ontological features of the mind are determined by analysis of cognition. The rejection of nativism is based on a view of how cognition and knowledge are to be possible. Introducing the imagistic theory of thought follows from the fact that it helps us find a direct link between thoughts and perception, grounding our thoughts and cognition in perceptual experience, thanks to which, in short, we can keep our thoughts in touch with the empirical world.[3]

Why is addressing the problem of grounding thoughts in perception epistemologically important? Linking perception and thoughts seems to provide a plausible naturalistic explanation of the genesis of our thoughts. It helps to clarify the ontogenesis and phylogenesis of our thoughts. Hence, linking perception and thoughts should provide a plausible description of how our thoughts stem from perceptual experience and what constitutes this possibility, that is, what kind of conditions one would have to accept if this link were possible. Burge (2010) and Prinz (2002) are, among others, prominent representatives of such an approach.

The great advantage of the imagistic theory of thought is that it is relatively straightforward to imagine how thoughts could be inherently perceptual, deriving directly from experience. From a general perspective, there is relatively little difference in how we characterize this link, whether it is based on the abstraction of simple sensations to more abstract ideas (e.g. Hume, 1977; Locke, 1975) or on a common perceptual symbol system (Barsalou, 1999). What is essential is that the imagistic model of thought makes this link conceptually possible, or, to put it differently, it would be more difficult to understand how any non-perceptual model of thought could explain it.

The problem of binding thoughts and perception is based on how one reconciles the categorically different features of two different kinds of mental acts. It is believed that perceptual representations have iconic content and format while thoughts are discursive. The high hopes put on the imagistic theory are related to the fact that it introduces a link between thoughts and perception. In the most general sense, the role of an image is to mediate between perception and thinking, being both the product of perception and the basis for thoughts at the same time. Playing a mediatory role means here, among other things, that images share features that can be ascribed to both perceptions and thoughts.

The mediatory role of images is usually understood as the claim that it is very likely that conceptual tasks, such as concept acquisition, categorization and inference, rely on perceptual representations (e.g. Goldstone and Barsalou, 1998). It is supported empirically by the body of evidence produced mainly in the 1990s and 2000s (e.g. Kan et al., 2003; Kiefer and Pulvermüller, 2012; Pecher et al., 2004).[4]

Without going into details, these results suggest that there is a link, presumably iconic, between thoughts and perception. It builds what Machery (2016) calls the 'neo-empiricist consensus', which united philosophers, psychologists and neuroscientists at the beginning of the twenty-first century (e.g. Barsalou, 2010; Martin, 2007; Prinz, 2002). A significant advantage of the imagistic theory of thought seems to be that it provides a reasonable picture of the nature of the link. Images seem to mediate between thoughts and perceptions, sharing features of both; thoughts are grounded in perception in virtue of being mediated by imagistic representation.

Thoughts–perception border

Although it seems as if the claim that thoughts are perceptually grounded provides a convincing justification for the imagistic theory of thought, it appears to be misleading upon closer inspection. For one thing, it blurs the distinction between thoughts and perception. For another, it appears that we overestimate the credibility of the case that seems to support the claim.

First, perceptions are not reducible to thoughts. Early perception is believed to be mostly[5] cognitively impenetrable and possesses a different kind of content than thoughts (Firestone and Scholl, 2016; Pylyshyn, 2003a; Raftopoulos, 2009; Raftopoulos and Zeimbekis, 2015). Moreover, the claim that thoughts and perceptions share a common representation format is insufficiently supported by empirical evidence. A vast number of studies show that thoughts, in contrast to perceptions, are amodal.[6]

Second, the idea that the content of a thought is perceptually grounded may be understood as the claim that every concept has a direct counterpart in an associated image that grounds the concept's meaning. The idea, however, faces obvious objections. Many concepts do not stand for any kind of image; for example, we do not need images for concepts like AND, NOT, YESTERDAY, TOMORROW, INFLATION, PRIME, INFINITY or NEGATION to understand their meaning. It is even difficult to imagine how we could have an image for empirical concepts such as PHOTON. And even if we had such images, it would be hard to determine what they mean. Mental images are private and cannot be the basis of public concepts. Thus, the Lockean account of thinking cannot provide a comprehensive theory of thought and was rightly rejected in the middle of the eighteenth century.

And here is the dilemma: although we have strong reasons to believe that thoughts are perceptually grounded, we have to admit that we have equally strong reasons to believe that we cannot interpret the nature of thought in purely perceptual terms. Perception and thinking are distinct mental capacities. It may be true that the border between thoughts and perception is not easy to determine, and there are many controversies about where the border is (e.g. Clark, 2013). Yet it does not follow that there is no border at all (e.g. Green, 2020). If all of that holds, then it may be doubted whether images may be common ground for perception and thoughts since that very commonality is logically excluded. Images cannot put perception and thought together like some other kind of representation cannot put colours and numbers together.

The difficult task at hand is to understand how to reconcile the claim that thoughts are perceptually grounded with the thesis that there is a border between thoughts and perception. Imagistic thoughts are believed to cover both cases. Therefore, they are supposed to possess characteristics that would enable us to interpret them as perceptual but not as instantiations of the genera of perception. Thus, any neo-Lockean explanation of what thinking with images is would have to be general enough to spell out what connects imagistic thoughts with perception and specific enough to point out what differentiates them. To make it clear, it is not logically excluded that it can be done.

Moreover, it is the case that every theory of thinking has to face this dilemma. The point is that it *cannot* be done simply by invoking a third term – an image – that mediates between perception and thoughts. Let me illustrate this problem with two contrasting neo-Lockean approaches: Jesse Prinz's (2002) theory of perceptual concepts and Christopher Gauker's (2011) theory of imagistic thinking.

The Lockean way of thinking about the perceptual grounding of thoughts was based on the idea that thoughts are abstract objects with an image-like nature. It may mean that the meaning of mental representations is grounded in relation to a corresponding mental image. However, such an interpretation is nonsensical, and it seems to have the same scientific status as the geocentric theory in astrophysics.

A more sophisticated way to spell out the neo-Lockean theory is based on the idea that mental representations do not require an imagistic counterpart to ground their meanings but have an image-like nature. Let me illustrate this concept with Prinz's proxytype theory.

According to Prinz (2002, 149), proxytypes are 'mental representations of categories that are or can be activated in working memory'. The central idea is that thinking is a simulation of perception – it is an idea borrowed from Larry Barsalou's seminal paper on perceptual symbol systems (1999) – and to think about something is to put oneself into a state that resembles the state of perceiving it. For example, the concept DOG is a set of perceptual representations acquired when one encounters, imagines or hears stories about dogs. All the experiences of dogs one has met, all of the images of dogs one has seen, all the stories one has heard, plus mental links that bind together those states build the concept DOG. On a given occasion, specific subsets of the set can be recruited to recognize or classify items in the world and form parts of thoughts as parts of larger imagistic or perceptual scenarios – these are proxytypes. They are context-sensitive. For example, if I think about a watchdog, I will invoke a representation of a German Shepard Dog and not a Poodle, but if I think about a cute small dog, I will invoke an image of a Poodle and not a German Shepard Dog. The 'proxytype' we construct on a given occasion depends on our cognitive needs in the moment.

The main advantage of the proxytype theory is that it helps us make sense of concept empiricism – the idea that all concepts are grounded in perception. It also explains some phenomena in cognitive psychology. For example, it explains the retrospective nature of representations that underlie many of the violation-of-expectance looking-time experiments or lags in understanding that manifest themselves in infant looking-time studies and tasks requiring explicit imagistic representations (e.g. Carey, 2009).

However, it is doubtful whether the theory of proxytypes solves the binding thoughts and perception problem, for the theory does not meet the Semantical Challenge. Notice that when one invokes two perceptual representations – a representation of a German Shepard Dog and a Poodle – we have two different representations. They do not represent what they have in common. The concept DOG does that. If concepts were perceptual representations of some general kind, such as a perceptual representation of a canine, we would need another perceptual representation that would determine the similarity between the representation of a canine, the German Shepard Dog, and the Poodle, and it would not stop there. We would need another representation, ad infinitum.

Moreover, a concept like the concept dog has a place in propositional structure. By substituting a representation of an object into an argument, we can form a proposition, such as ALL DOGS ARE MAMMALS. If the concept DOG were a perceptual representation of the same kind as other perceptual representations, such as a representation of a mammal, there would be no way to determine whether the representation DOG plays the role of a named argument or a predicate in a proposition. In other words, if mammals and DOG were both perceptual representations, it would be impossible to distinguish between two different propositions like ALL DOGS ARE MAMMALS and ALL MAMMALS ARE DOGS. Prinz recognizes the problem and tries to develop the idea of mental links that bind together perceptual representations and determine the functional roles of perceptual representations. But what he achieves is introducing concepts under a different name, implicitly restating that perceptual representations alone cannot be concepts.

Gauker (2011, 2017) approaches perception and concepts differently. Prinz posits that there is a continuity between thoughts and perception because concepts are perception-like. However, it contradicts the claim that there is a border between perception and thoughts. Gauker's idea is that there is continuity and a border between thoughts and perception, for thinking can be divided into two kinds. The first kind of thinking is conceptual and language-dependent. It introduces the border between perception and thoughts. The second kind of thinking is non-conceptual and imagistic – it introduces the continuity between perception and thoughts.

The general idea is that perceptual representations cannot be the basis of concepts and abstract representations. Perception of a table is always a perception of a particular object and cannot represent kinds such as A PIECE OF WOODEN FURNITURE or TABLE – language can. Gauker's alternative is to demonstrate that thinking can consist in non-conceptual operations based on mental images. To explain the nature of imagistic thinking, Gauker holds that we can see imagistic thoughts as regions of perceptual similarity spaces and as topological maps of objects in the brain.

Imagistic thinking is described as a capacity to locate and track objects and scenarios in perceptual similarity space. A perceptual similarity space is a topological hyperspace in which the dimensions represent various ways in which objects may perceptually differ. A mark in a perceptual similarity space is a vector of the perceptual similarity space. The location of a mark on a dimension of the perceptual similarity space is the mind's measure of the location of a sensory object or a scenario on the dimension of the objective quality space that corresponds to it. For example, when the mark representing the German Shepard Dog is closer in perceptual similarity space to the mark representing the Poodle than a Persian cat, then a subject assesses that the German Shepard Dog is more similar to the Poodle than to a Persian cat. The general idea is that similarity spaces can explain how we recognize and group objects in perception without postulating language-driven concepts.

Gauker's idea is that we can think about thinking not in terms of operations on concepts but in terms of perceptual similarity spaces and imagistic cognition. According to Gauker, concepts are a matter of mastering language. Thus, concepts are building blocks of language-like propositions. However, not all thoughts are conceptual. The medium of conceptual thought is language. The medium of non-conceptual thought is an image (Gauker, 2007). To represent an object imagistically is to assign it to a location in the perceptual similarity space.

However, there are two problems with Gauker's non-conceptual view of imagistic cognition. First, the problem concerns a transfer between conceptual and non-conceptual representations. Imagistic thoughts cannot be easily put into words, but that does not mean that images cannot be put into words at all. For instance, I can imagine a cat sitting on a mat, expressing my ability to recognize objects in perceptual similarity space, but I can also form a definite description 'the cat sitting on a mat'.

The problem is that we want to maintain the possibility of translation between perceptual and conceptual thoughts. However, such a possibility is excluded since they share different metaphysical properties. According to the Metaphysical Challenge, conceptual thoughts are compositional and systematic; imagistic thoughts are not. Translating a non-conceptual system of representation into a conceptual one is not like translating German into English but like trying to put a set of randomly generated phonemes into English.

Second, one of the main advantages of Gauker's theory of non-conceptual imagistic thought is its capacity to explain the behaviour of non-linguistic animals, such as non-human animals and infants. Imagistic thinking seems to be more primitive and more fundamental than language and language-based knowledge. However, the problem arises when we try to extend the thesis of the non-conceptual content of imagistic thinking onto the domain of scientific knowledge. Pictures can be a source of knowledge, as scientific practice shows. We can ask, however, what is the nature of such knowledge. It could be tempting (e.g. Mößner, 2018) to argue that pictures and visual perception are bearers of non-conceptual knowledge and can show something that is not yet conceptualized. So far, we seem to align with Gauker's theory of non-conceptual content. Yet, as the Epistemological Challenge shows, it cannot be the case. Science consists of theories and hypotheses that are built from propositions and concepts. The goal of science is to put things in the conceptual form of theories and hypotheses. Any content that has not yet been conceptualized cannot be part of scientific knowledge.

Thus, it seems that neo-Lockean strategies fail to explain the thought-perception problem. The proxytype theory does not meet the Semantical Challenge. Gauker's theory of imagistic cognition does not meet the Epistemological and Metaphysical Challenges. It does not mean that the thought-perception problem has disappeared. Even though we do not have a satisfactory solution – and because of that – we still need one. The point is that the problem cannot be easily solved by invoking imagistic representations as theoretical concepts that can make this problem solvable. The problem is even more profound. It is controversial whether postulating the existence of images as mental objects we think with makes any sense.

Wittgenstein–Ryle's sceptical argument

The main philosophical problem with the question 'What is thinking with images?' is connected to the overwhelming temptation of explaining away the problem, which is based on the assumption that the very question is ill-posed. As one may argue, it is not the case that we think with images, just as it is not the case that we think with or in words.

This claim is most frequently expressed by the view that we think neither in images nor in words but in *thoughts*. Although such an answer to the question may seem tempting, it is nonsensical. To say that we think in thoughts is no more informative than saying that we do not speak in French or English but use language. The misleading character of the answer follows from the fact that to study language, we study particular instantiations of language – for instance, French or English. The same is true in the case of words and images. We can only study the nature of thoughts by studying particular instantiations of thought processes expressed in language and images.

A more sophisticated way to render the question absurd is to argue that the very expression 'thinking with' is senseless. Expressions such as 'thinking with words' or 'thinking with images' lack meaning. Wittgenstein (1953, 1958) and Ryle (1979) undertook this sceptical line of argument (Bennett and Hacker, 2003; Slezak, 2002b).

Let me illustrate this strategy with Ryle's (inspired by Wittgenstein) argument against the belief that thoughts consist of word-like objects that are expressed in language. According to Ryle (1979), it would be senseless to say that one composes their speech expressing some English-like thoughts. Considering English words and phrases, modifying and rejecting others, stringing together words into sentences and paragraphs, one is not thinking in some English-like medium. Their speech will be, of course, in English, but we cannot say that the mental process of composing the speech was in these words. They did not have these words while composing the speech. They were searching for these words in the process of speech composition. Thus, the expression 'thinking with words' is senseless.

The general objective of a Wittgenstein–Ryle style argument is to undermine a dominant view on the nature of thinking, which could be put in terms of a vehicle-cargo model. According to the model, thoughts consist of the content of the thought – something the thought is about – and the bearer of the thought – something the thinking takes place in, for example, in words, in images or any other representational system. According to Wittgenstein and Ryle, this model leads to nonsensical consequences and should be abandoned.

As clearly stated by Bennett and Hacker (2003), the source of the confusion is the idea that we think *in* a medium, just as we speak *in* a language. We can even say that we think in English, and not in German, and if not in English, then in French. And even if we do not think in a language, we must think in something else, for example, in images. But that is only a conceptual confusion.

Talking to oneself is neither necessary nor sufficient for thinking. I can talk to myself without thinking, for example, when counting sheep, to *stop* thinking and fall asleep. I can think without talking; for example, I can directly formulate a conclusion based on some evidence, without saying that p follows from q, if r, and so on. It does not mean that thoughts are primarily in images, being a basis for translation from images into words. To use a word to express a thought has nothing to do with having an associated mental image. According to the private-language argument, if having an associated mental image were necessary for understanding the meaning of a word, then it would be impossible to determine the meaning of the word. Thus, one need not think in anything; particularly, one need not think in images.

Moreover, the sceptical line of argument could even be strengthened by pointing out that whether we think in language or images is misleading in two different ways. One can argue that we do not have to think in language but must possess the language to express thoughts. As an example, we need language to express a rational train of thought. Note that there are only thoughts that could be (but do not need to be) expressed. The concept of an inexpressible thought is in the same sense senseless as a function that does nothing. Therefore, we have to master the language to be able to think rationally. Thus, the words one uses when one thinks are the expressions of one's thought.

In contrast, images are neither necessary nor sufficient for expressing a rational train of thought. Images may cross one's mind while reasoning, but they can only play the role of heuristic devices to help one find a solution. They can aid thought as an accompaniment of thinking, but they are not thoughts or expressions of thoughts.

However, we should not feel intimidated by these arguments. Following Goodman (1984), there is a notable linguistic difference between words, images and other objects we can think of. Linguistically speaking, it seems acceptable to say that we can think *in* words and *in* images, as well as *of* words and images. In contrast, we can think *of* cabbages, but we cannot think *in* cabbages. Analogously, we can speak in words without speaking of words, and we can speak of cabbages but not in cabbages. What does it mean to think in words or in images?

First of all, it cannot be reduced to thinking of words or images, just as worship with images is not reducible to the worship of images. Thinking of cabbage is not reducible to thinking of a word or an image representing a cabbage since the thought of a word or an image would need another word or image, and so on, ad infinitum – that was Ryle's primary point. Words and images are somehow in mind, while the cabbages, words and images we think of are not.

Second of all, thinking of words and images and thinking of cabbages do not require producing them. The phrases 'thinking in words' and 'thinking in images' do not mean that thinking is a three-term operation between thinking, the object of thought and some entity, a word or an image, that moderates between the act and the object of thought.

In contrast to the Lockean approach, images do not have to be postulated mental objects. I can think of a hexagon and draw a hexagon without forming an image of a hexagon in my mind. Such a proposal does not require presupposing objects such as words and images we think of to substantiate thinking in words and images. Instead, thinking like speaking and manipulating images may be interpreted, in Kantian and Fregean fashion, as an operation conducted in the mind, where mental and extramental images may be interpreted as the final results of those operations and necessary means to represent these operations.

The term 'operation' is polysemic. In general, it means a rule-governed process of taking input values into the output values. Notably, operations are not limited to computations. Instantiations of operations are subtraction, construction, functions mapping propositional contents onto a set of truth values, a partial ordering function and so on. Two points have to be stressed. First, the distinction between operations and the results of operations is not the same as the distinction between actions and the products of actions (Twardowski, 1999). To be an operation is to be a rule-governed process, for example, a function, regardless of whether the operation is carried out in a relevant action.

Second, operations and the results of operations are different sides of the same coin. Every operation can be reformulated into a relevant result of an operation, and vice versa. For example, mathematical axioms can be reformulated into inference rules and the other way round. From a logical point of view, their status is equivalent (e.g. Stegmüller, 1969).

Therefore, we have to distinguish between images as results of operations and image-producing operations. The claim that there is no image in mind mediating the process of thinking in images does not mean that there is no operation that produces an image. Consequently, when we inquire into the nature of thinking in images, we are not inquiring into the nature of objects such as images in mind but the nature of image-producing operations, resulting in relevant images. In other words, our questions are about the nature of a family of operations labelled 'thinking with images'.

The operational approach

Interpreting the relation 'thinking with' in terms of operations helps us better understand the nature of the concept of a 'bearer of thought'. In short, the notion of a bearer of thought is interpreted here in terms of a family of operations conducted in a representable medium. The basic idea is that thoughts are operations that need certain vehicles to be carried out and represented, such as words or images. In the same way, counting is a function that to be carried out requires that we introduce certain vehicles, that is, numbers we count with and numerals that make those numbers representable.

Importantly, numbers do not exist outside of those operations, for example, outside of the context of counting. If one asks what numbers are, one does not ask about some objects in Platonic heaven. One asks what numbers do, that is, what kind of operations they can be part of. For example, if one asks what numbers are, a valid answer may be 'something we count with to determine a less-than relation between a and b'. At the same time, numbers can be seen as the final results of these operations. For example, one can say that the number 2 is the final result of an operation determining the cardinality of a set; that is, it is a measure of the set's size. To be the number 2 is to be more than 1, 0 and less than 3, 4 and so forth. Searching for numbers as ideal Platonic objects is misguided.

Moreover, numbers require numerals to be represented, for the limits of possible operations are the limits of possible representations of the operation. It means that every number ought to be representable, though not necessarily actually represented, by a relevant numeral. An unrepresentable operation does not exist, for the nature of an operation is that it does something. It can do something only in some representable medium, such as words, images and numerals. An operation that does nothing is not an invalid operation. It is not an operation at all.

The point is that just as there is no counting without numbers and without numerals that represent them, images can be interpreted as necessary vehicles of thoughts. In the same sense, some operations would be impossible to carry out if no images instantiated and represented the operations. The nature of counting and thinking with images is determined by the nature of the operations, not by the nature of the vehicle, such as the features of numerals and pictures. Numbers can be represented by digits, words, such as 'one' and 'two', and apples. What determines that an expression is a numeral is not that it consists of words or digits but the fact of what the expression represents. Every representation of a number is a numeral, whether formed in words or apples. To understand the nature of numbers, one must understand what kind of operation it is a part of. For example, to understand the number 2, one must understand that $2 + 2 = 4$ or that a set contains two and only two elements.

By the same token, to determine what it is to count and what it is to think with images is to determine what we do, not what numerals or pictures hanging on a wall are. At the same time, numbers and images are not tools to think with, where the nature of the operation of using a tool is separable from the features of the tool. Numbers are not like calculators, for it is not the case that when learning how to count, one first learns how to perform calculations and then what numbers are. Numbers are necessary if operations like counting are to be carried out. Similarly, images are necessary for thinking in the sense that some operations would not be possible without them. For instance, it would be nonsense to say that we first learn how to form an image, such as a drawing, then try to find tools to draw. Learning how to draw is inseparable from drawing.

According to the operational approach, imagistic thoughts are like numbers. They are defined by the operations they are a necessary part of. Just like numbers, imagistic thoughts require a representable medium, that is, images, to express these operations. Consequently, the question 'What is thinking with images?' is interpreted as equivalent to the question 'What kind of operation is expressed by images?' The question may be confusing, just as it may be confusing to ask 'What kind of operation is expressed by numerals?', but that does not mean that it is senseless. And even if the operations we are looking for are not the same operations as the ones expressed in language, no one has to assume that they have to be the same. Not all thoughts can be expressed in language; some can only be expressed in images. The point is to find out what these thoughts are.

In contrast to the neo-Lockean approach, the operational view is non-committal on the point of any mental objects and episodes displayed before the mind's eye, such as words or images we think of. The intuitive idea is that when we think of, for instance, what the result of an operation like 2 + 2 is, we do not have to write down a formula on a piece of paper or construct an explicit mental representation consisting of the symbols '2' and '+'. We think in numbers without having to think of them. However, any operation has to be conducted in a representable medium, for example, in words and images.

Thinking about images in terms of operations may seem counter-intuitive. A comparison to the propositional theory of meaning in the philosophy of language offers a better picture of the point I am presenting.

According to the propositional view in the philosophy of language, the meaning of the sentence S is the proposition P expressed by S. Much ink has been spilled to argue that the propositional view is incomprehensible. For example, we know neither how propositions exist nor whether they exist in time and space. As a result, it could be argued in a Quinean (e.g. Quine, 1960) fashion that we have to reject the propositional view as postulating ontologically unnecessary entities.

However, the propositional view is non-committal on the point of the existence of abstract entities such as propositions. For instance, intensional semantics interprets propositions in terms of functions from possible worlds to truth values. It means that propositions are operations that determine the conditions according to which a sentence S is true or false in a possible world. The indispensable role of S is to make this function representable. Therefore, we can accept propositions as bearers of meaning without positing unnecessary abstract objects.

The main point is that we can accept images and words similarly as bearers of thoughts without being committed to the existence of image- or word-like objects in mind. Images and words express some meaningful operations and are indispensable for those operations to exist. The concept of a bearer of thought is no more mysterious than the concept of a proposition, and therefore Wittgenstein–Ryle's argument misses the mark. The question of the nature and function of imagistic representations in thought processes is the question of the nature and function of bearers of thought in the same way as the question of the nature of the meaning of words can be interpreted in terms of the question of the nature of propositions. And although that does not mean that the question of the nature of propositions is easy, it is not senseless.

One caveat is in order, the operational view is not reducible to the dispositional view, and thinking with images is not reducible to a disposition to produce a mental image (Price, 1953), to a certain skill (Bennett and Hacker, 2003; Ryle, 1949) or the state of readiness for producing images and for judging presented images as agreeing with one in mind (Goodman, 1984). To put it more precisely, interpreting the nature of an imagistic bearer of thought in terms of a disposition does not suffice to explain the operations of using images in thinking.

Granted, possessing a disposition to produce and interpret an image is necessary for thinking with images. In the same way, to understand the meaning of a knight in chess is to have a disposition or a skill – interpreted loosely as a regular disposition – to move it in a certain way.

However, it is not a sufficient condition. One can be disposed to think and speak like Kant without any understanding of Kantian thought. One can have the disposition to recognize similar pictures of a triangle without understanding that the pictures represent a triangle. To master an operation, one has to know what the correctness conditions of the operation are. In other words, one has to know what counts as a correct instantiation of an operation.

In the same way, mastering the counting operation is to master a certain skill concerning counting, but explaining the nature of counting is not the same as explaining the nature of the corresponding disposition. Explaining the nature of counting is to explain the rules governing a family of functions. An explanation, such as a psychological explanation of the nature of the disposition, may be part of an explanation of a function, yet it is not reducible to the latter.

To sum up, to make sense of the question 'What is thinking with images?', it shall be interpreted here as a question concerning the nature of operations expressed by images. It shall *not* be interpreted as a question concerning the nature of some mental objects, such as imagistic thoughts, picture-like ideas or mental images, or a question about the nature of dispositions. In short, the images we think with are primarily something we *do* and not something we *possess*.

Consequently, we can infer the nature of the given operation from studying external instantiations of the operation. In the same way, we investigate the nature of abstract objects, such as functions, by examining their instantiations, for example, equations, and the nature of mental phenomena, such as linguistic meaning, by studying linguistic representations.

Explaining operations

As I have argued, any attempt to answer the question 'What is thinking with images?' has to begin with careful isolation of what is to be explained and an understanding of what kind of answers may be considered satisfactory. For in the case of such operations as thinking, the very concept of explanation may lead to confusion. The issue is particularly prominent in explanations of the nature of such objects as gold and water.

When we try to explain the nature of gold, what is to be explained is the nature of an object – a sort of yellow metal – and the *explanans* is the structure of gold and the relations of gold to other elements. Let us say that gold is the chemical element with the atomic number 79. We communicate the number of protons in the element and the element's position in the periodic table. If we say that water is H_2O, we are trying to explain the nature of some wet substance. Our explanation is a description of the chemical structure of water and the relation of water to other elements, for example, to oxygen.

In the case of thinking, the situation is different, for there is no thinking-like object that is to be explained. As Ryle noted (2009, 271):

> When we start to theorise about thinking, we naturally hanker to follow the chemist's example, namely, to say what thinking consists of and how the ingredients of which it consists are combined. Processes like perspiring, digesting, counting and apple-picking can be broken down into ingredient processes, coordinated in certain ways. So it seems reasonable to expect thinking to yield to the same treatment. But this is a mistake. There is no general answer to the question 'What does thinking consist of?'

In contrast to the object-like view, we can think about thinking as an operation, in the same way as counting can be described in terms of a mathematical operation, meaning in terms of mapping onto a set of possible worlds and language as a set of semantic and syntactic rules. The result of thinking is a thought, in the same way as the result of using a language might be a meaningful sentence, and the outcome of a calculation might be a formula written on a piece of paper. However, if we want to explain what thinking, language and calculation are, the object of explanation is not some thought-, language- or counting-like object. Suppose we are to understand what thinking, meaning, language and calculation are. In that case, we must understand the nature of the operations expressed by certain utterances, sentences and formulas. Therefore, explaining the nature of thinking is explaining the nature of an operation. What does it mean to explain an operation?

Notice that if one wants to explain the nature of an operation, such as calculation, what is to be explained is not describable in terms of the phenomenology of calculation. In that context, phenomenology is irrelevant; that is, a phenomenological description of calculation leads us nowhere if, by phenomenological description, one understands a description of symbols and formulas used in the act of calculation. For the nature of the calculation, it is irrelevant, for example, whether we use the symbol '1000' or '10^3'. It is also irrelevant whether the calculation is done with digits or words. What distinguishes calculation from the description is not the use of digits, on the one hand, and words, on the other. The term $2 + 2 = 4$ can be written down as 'two plus two equals four'. What matters is the operation that is performed, which means that to isolate the object to be explained, we have to separate the relevant logical relations understood here in terms of a family of functions that are carried out on a certain set.

If the 'family of functions' is an explanandum of the nature of the calculation, what is the explanans? Notice that if we want to explain the nature of the 'family of functions', we do not relate it to another, presumably more complex and detailed family of functions. For example, trying to explain what function PLUS is, we do not say that it consists of some more detailed parts, such as function $PLUS_1$ and $PLUS_2$. In contrast, if we interpret the formula 'a chemical element with the atomic number 79' as an explanandum, the explanans could describe a more fundamental structure. It could be put, for example, in terms of a structure consisting of quarks and leptons that are parts building larger structures. In other words, a more profound explanation of something such as gold is a description of a more detailed structure.

Understanding the nature of an operation, such as calculation, is quite different. In this case, we need to understand the conditions that have to be met for the operation to be carried out. For example, if we want to understand what the operation PLUS is, we may point out instantiations of the function PLUS and some set-theoretic assumptions that have to be accepted if the function is to be logically possible. In this case, for example, we may adopt Zermelo-Fraenkel's axiomatic system. A more profound explanation of the operation would relate to a richer conceptual scheme, for instance, to some second-order logic or category theory axioms. Similarly, if we want to explain the interpretability of language, we may adopt compositional semantics. Compositionality would function here as part of the *explanans* of the interpretation function in an argument of the following form: if language is to be interpretable, which is the case, then it has to possess a compositional structure.

The same conclusion applies to the case of thinking, in general, and thinking with images, in particular. First, to explain the nature of thinking with images, one would have to designate an explanandum, isolating the functional roles of images in thinking. In other words, the first objective is to establish what images do in the operation of thinking.

Second, after isolating an explanandum, the required explanation needs to describe the conditions that must be met if images can perform certain functional roles.[7] Therefore, the first step is to investigate the functions of images in cognitive processes to determine which functions are in accord with the imagistic theory of thought. The point is to distinguish such operations that cannot be carried out without images and distinguish them from non-imagistic ones, such as operations requiring numerals and words to be expressed. The second step is to establish conditions that have to be met if images are to cover the requirements of the latter. It has to be established how the concept of an image needs to be interpreted if images can play the designated role. In short, to answer the question 'What is thinking with images?', we must first ask what thinking with images does and then we must try to find something that does that.[8]

The Traditional View holds that the answer to both questions is that images represent by resembling objects. I will show that such an answer is not completely wrong but not completely right either. In the following chapters, I show that images are essential for the operations of constructing and recognizing objects. Following that, I show that an image can represent the ways of constructing and recognizing objects by exemplifying these ways.

Summary

To answer the question 'What is thinking with images', one has to determine what kind of answers one expects. There are two strategies available. Neo-Lockean theories hold that imagistic thoughts are theoretical objects introduced to explain the relation between thoughts and perception. Imagistic thoughts are abstract entities mediating between perceptual and discursive representations.

However, the problem of the relation between thoughts and perception cannot be solved by invoking the concept of imagistic representation. There is a thoughts–perception border, and no intermediary object can bring together representations that possess mutually exclusive characteristics.

In contrast to Locke, Wittgenstein–Ryle's sceptical argument holds that thoughts cannot be kinds of objects. In particular, there is no vehicle we think with; thus, the question is senseless. If there are no objects we think with, the question of what characterizes thinking with images is deprived of content.

I argue that Wittgenstein–Ryle's strategy is an attempt to explain away the problem, not to solve it. I introduce an alternative strategy, according to which we can understand the subject of the question in terms of operations based on imagistic representations. It means that the question 'What is thinking with images?' is equivalent to the question 'What kind of operations are expressed by images?' Accordingly, I present a way to formulate a sensical and non-trivial answer to it. It is put in terms of the operational approach, which means that thinking is a matter of carrying out certain operations and explaining the operations is equal to pointing out the conditions that have to be met if such operations are to be possible. In a nutshell, I argue that to answer the question 'What is thinking with images?', one has to find out what images do and then find something that does that.

4

What do images do?

According to the operational approach, to understand what thinking with images is, one has to understand what kind of operations images can perform and find something that does it. The traditional answer to this question is that images represent objects by means of resembling them. However, left as it is, this answer is insufficient, for it does not address the issue of the irreducible role of images in thinking. Moreover, it makes imagistic thoughts vulnerable to epistemological, semantical and metaphysical objections described in Chapter 2.

In the following three chapters, I explore the irreducible roles of images, looking at two case studies taken from mathematics and physics: a diagram of a knot and a picture of a black hole. First, I argue that the way they represent the world cannot be explained by resemblance-based and description-based semantic theories; second, they demonstrate genuine semantical functions of images. I hold that images represent the ways of constructing and recognizing objects. I explain how to understand the operations of construction and recognition, and, drawing on that, I introduce the two-dimensional model of iconic reference and the measurement-theoretic account of images. In a nutshell, I hold that images exemplify the rules of construction that enable us to identify the depicted objects. The best explanation is that images should be taken as measurement devices comparable to rulers and balances.

Case Study 1: Knot diagrams

Drawing on the case of knot diagrams, I demonstrate that one of the roles of images is to represent the ways objects and events can be constructed. Let me start by describing the knot theory.

Knot theory is a branch of topology dealing with knots and knot diagrams. A knot is a smooth closed curve in a three-dimensional Euclidean space. For every knot, there exists a knot diagram representing it (Cromwell, 2004, Theorem 3.3.2). A knot diagram is a representation of a knot with the height information at the intersection points, which tells us which strand passes over the other. An unknot is a knot representable by a knot diagram without crossings. A trivial example of an unknot is a ring.

Knots can be transformed in various ways and still remain the same knot, so long as they are not cut. Properties that hold throughout such deformations are called invariants. Two knots are equivalent when one knot can be smoothly transformed into the other without cutting, gluing together or allowing a self-intersection during the transformation. The equivalence relation captures the intuition that some topological properties of objects remain unchanged under deformations, regardless of the changes in the geometric shape of the object. The main task of the knot theory is to determine the equivalence of knots, as represented in Figure 4.1.

One way to determine the equivalence of knots is to demonstrate that two knot diagrams can be transformed into each other by a sequence of three kinds of moves, known as Reidemeister moves. According to Reidemeister's Theorem, two knots are equivalent if and only if there is a sequence of Reidemeister moves transforming a knot diagram of one knot into the knot diagram of the other. This means that for any two projections of the same knot, a sequence of Reidemeister moves will transform one projection into another.

There are three types of Reidemeister moves (Figure 4.2). Type I allows us to twist the knot or remove one while the rest of the knot remains unchanged. Type II allows us to move one loop over another. Finally, type three allows us to move a string over or under a crossing.

Applying Reidemeister moves is sufficient to determine the equivalence of knots. For instance, we can demonstrate that the knots represented in Figure 4.1 are equivalent by applying the Reidemeister move of the first type to both of

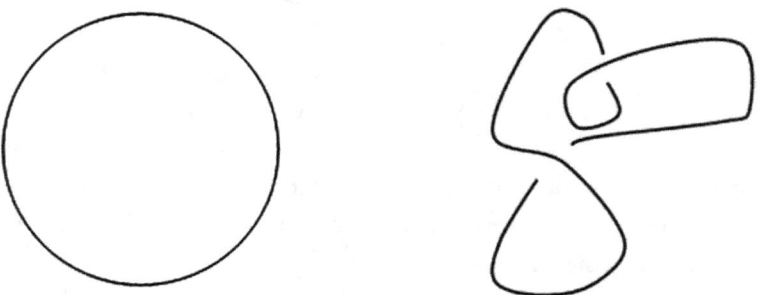

Figure 4.1 The unknot (on the left) and an equivalent knot (on the right). © Piotr Kozak.

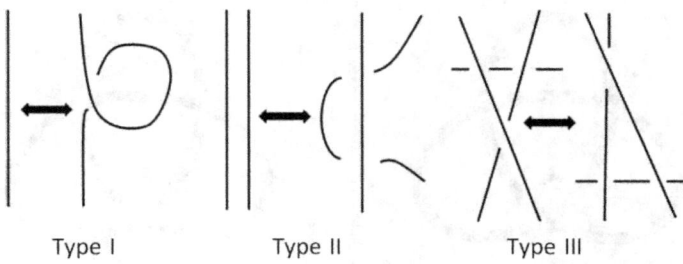

Type I   Type II   Type III

Figure 4.2 Reidemeister moves. © Piotr Kozak.

the crossings. However, there is no known algorithm for applying Reidemeister moves. It is just trial and error.

Computational methods in knot theory are based on the concept of knot invariants. A knot invariant is a function that assigns a certain algebraic object to each diagram of the knot so that the same values correspond to equivalent diagrams. If we can find an invariant that takes different values for given diagrams, then the diagrams are not equivalent. There are computational notations of knots, such as the Dowker notation or the Conway notation, that help to solve the problem of knot equivalence by applying the algebraic description of the knot.

Does this prove that algebraic ones can replace image-based methods? Definitely not. Suppose that we can provide sufficient equations to solve the problem of knots equivalence. Yet, what we want to know is not only whether one knot can be transformed into another but also *how* it is done. We want to know how to construct a knot, that is, what are the steps that lead from an unknot to a knot. Metaphorically speaking, even if we know that a tie can be tied, we still want to know how to do it.

Let me explain. Some formalizations can describe a diagram of a knot. For instance, the Dowker notation can capture the first and second Reidemeister moves. However, without the diagram, we cannot interpret these notations. Consider the following example. Dowker notation represents the first Reidemeister move by representing the number and character of the crossings with consecutive integers. However, we still need to know how to perform the first Reidemeister move. Without such knowledge, we would be left with a meaningless mathematical formalization. We have to know what kind of construction these numbers represent. Therefore, image-based operations are irreducible in knot theory even if we can provide formal descriptions of the knots.

To illustrate it, let us consider the trefoil knot. The difference between the left-trefoil and right-trefoil knots is well captured by two knot diagrams represented

Figure 4.3 The left-trefoil knot and the right-trefoil knot. Source: Wikimedia Commons.

in Figure 4.3. However, this difference is not representable by Conway polynomials. The Conway polynomial of each kind of the trefoil will be the same. To capture this difference, it is necessary to introduce Jones polynomials. In this case, diagrams are necessary to determine what constructions are representable by a mathematical description.[1]

Diagrams are necessary to represent knot constructions for they exemplify them (more on exemplification in Chapter 6). The theorem that every knot is representable by a diagram should be read as follows: I can understand the description of a knot construction only if I have the disposition to understand a diagram exemplifying this construction. In contrast, I can understand the diagram without knowing the corresponding mathematical description. For instance, I can understand the description of the first Reidemeister move only if I can identify the first Reidemeister move in a diagram. In contrast, I can identify the first Reidemeister move in the diagram, without knowing its description. Thus, understanding the description of constructions is built upon the ability to understand constructions in diagrams.

To sum up, a knot diagram manifests how knots can be constructed. It shows possible transformations of knots according to specific inferential procedures. This aspect of knot theory is not reducible to the problem of knot equivalence and can be represented by diagrams with well-defined operations, such as Reidemeister moves. In the following sections, I explicate the concept of construction and explain its implications for the theory of the content of diagrams.

What are constructions?

Let us reconsider the knot diagram represented in Figure 4.1. What distinguishes this diagram from a scribble is that it has content, that is, it can be correct and incorrect. However, there is no one predefined interpretation of this diagram. There are infinite ways to get from the knot to the unknot. It can be twisted, unfolded or its orientation changed, but it cannot be cut. What determines the correctness standards of these interpretations?

To understand what a knot diagram represents, one has to know what procedural steps are allowed by the diagram. De Toffoli and Giardino (2014) identify it with the dynamic nature of diagrams. A dynamic diagram is a representation closely related to specific inferential procedures that help define possible transformations of the diagram (De Toffoli and Giardino, 2014; Hegarty, 1992). Our interpretation of a knot diagram is based on the ability to imagine performing some operations on the diagram according to some inferential procedures.

However, not all diagrams have a dynamic nature. A diagram of a triangle is not dynamic. Moreover, we do not have to visualize transformations. We have to know what kind of transformations are correct and incorrect. Imagining transformations of the diagram can be a means to identify correct and incorrect transformations.

As I have argued, knot diagrams exemplify the ways objects can be constructed. I call these ways 'the rules of construction' of an object. Let us unpack this definition.

The concept of a rule is understood here as a function that applies correctness standards to some construction procedures. Construction is correct if it is in accord with a rule. Otherwise, it is incorrect. A rule describes some moves as correct or incorrect, depending on whether they are allowed by the rule. Importantly, knowing a rule means neither knowing its description nor the propositional knowledge-that, such as knowing the content of an algorithm. It is a kind of procedural knowledge that determines whether a move is correct or incorrect.

The knot diagram in Figure 4.1 represents the rules of transforming a knot into an unknot by applying the Reidemeister move of the first type. The Reidemeister moves are transformations that are allowed by the construction rule, cutting any of the knot's crossings is not permissible. The diagram's content is interpreted based on recognizing these rules. In short, I know that the knot diagram represents the unknot since I understand how to transform one into another.

Rules of Construction: functions ascribing correctness standards to construction procedures.

How to understand the idea of exemplifying rules of construction? Consider the case of a tangram (Figure 4.4), a game of putting together flat figures to arrive at the desired pattern (Ertz, 2008).

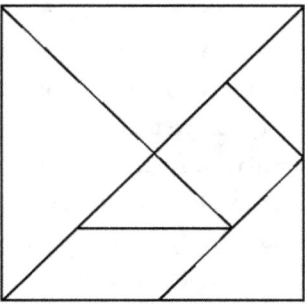

Figure 4.4 A square pattern matched with a tangram set. © Piotr Kozak.

Note that tangram rules are exemplified by an outcome of applying these rules represented in Figure 4.4. If someone asks how to play tangram, I can show him Figure 4.4. In the same way, if someone asks how to build a house, I can show him possible element arrangements in an architectural project. The way we construct an image is inseparable from the outcome of this construction displayed in the image. The image exemplifies the rules of construction.

Exemplification of construction rules is generalizable (Lehrer, 2000). Suppose that I am asked to describe tangram rules. I can point to a particular image exhibiting a pattern. Yet, this is not to say that I am looking for only this particular image or the images that correspond to its description. It is essential to understand how to go on. To understand the rules of construction, one has to possess non-propositional knowledge of how to apply these rules.

Let us compare it to the way a ruler works. A ruler exemplifies a measure by means of which you identify length. However, it does not mean that a one-metre-long ruler can locate only one metre. Knowing how to use a ruler implies that you can measure all distances.

There are two types of construction rules: the rules of interpretation and the rules of production (see Hyman, 2006). Rules of interpretation determine how to identify the represented object (more on the recognition-based identification in Chapter 5). Rules of production determine how to create the representation or the represented object. The ability to apply these rules is dissociated. I can understand what is being represented in a diagram without having the ability to create the diagram or the represented object. In the same way, I can understand a foreign language without being able to speak it.

It is easy to confuse these rules. However, it is even more important not to separate them. They are not different kinds of rules. They are different *applications* of the same rules of construction. They both determine the correct and incorrect moves to localize the represented object. In the same way, interpreting a sentence and producing the sentence are governed by the same rules of grammar. Even though these rules are applied differently, they can be easily brought together.

In the case of applying the rules of interpretation, it is essential to identify in a representation the information to localize the searched object. In the case of the rules of production, it is essential to identify the matching elements to create this object. Let us consider a ruler. It shows you how to find the searched distance and how to create a line segment. By the same token, a triangle diagram can show you how to form a triangle (according to the rules of production) and identify it (according to the rules of interpretation). In both cases, the diagram represents the rules of construction of a triangle.

What is construction? It is usually interpreted as a procedure (or an outcome of such procedures) of employing instruments, such as a ruler and a compass, to draw a geometrical figure. This is called classical construction. However, in the most general sense, employed by Euclid and Proclus, the concept refers to a procedure of adding something that is 'lacking in the given for finding what is sought' (Proclus, 1992, 159).[2] Let me elucidate it with the help of a Euclidean example.

Take three line segments a, b and c, which satisfy the condition that any two line segments together are longer than the third. This is 'the given' in Proclus's description. Next, we construct a line segment DE, and we place on it the line segment DF that corresponds to the length of a. Then we construct the line segment FG that corresponds to b. We do the same with the line segment GH that corresponds to c. Then we construct two circles. The first circle has a centre F and radius FD. The second circle has centre G and radius GH. The two circles intersect at K. It is now enough to join K, F and G to arrive at the construction of $\triangle FGK$ (Figure 4.5).

Knowing the length of a, b and c and the definition of a triangle is insufficient to solve the task of constructing a triangle. We need to know the rules of construction for triangles. Even if we seek a polygon with three edges and three vertices, we need to know the correct moves leading from the line segments a, b and c to a triangle. Knowing the construction rules allows you to arrive at a triangle if you are given three line segments and know what a triangle is. The diagram represents these construction rules.

The concept of construction is reducible to neither the concept of the construction elements nor the definition of the constructed object. It is possible to get different triangles from the same definition and the same line segments. In the case of Figure 4.5, you will get triangles with different properties, for

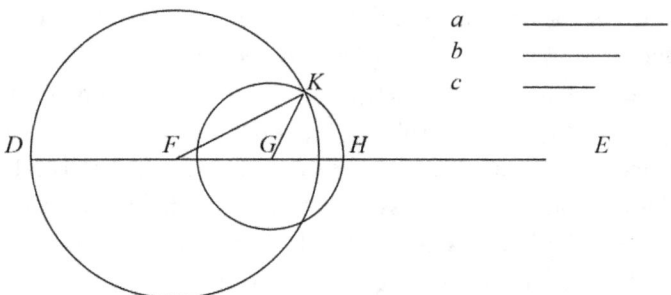

Figure 4.5 The construction of the triangle ΔFGK. © Piotr Kozak.

example, different heights, depending on which line segment is used as the base. Construction is a skilful application of these elements in order to get to the defined object.

According to Tichy (1986),[3] the concept of construction can be generalized and applied to non-geometrical objects. Following Tichy, I characterize constructions as denoting the procedures of arriving at a goal (an object, event or scenario). How does this apply to non-geometrical objects?

Let me use an algebraic example.[4] Note that the term '7 + 5' denotes the same mathematical object as the term '20–8', namely the number 12. However, these terms denote 12 in different ways. Apart from naming 12, they identify different procedures for arriving at 12.[5]

The construction '20–8' consists of '20', '8' and the subtraction function. In the same way, the construction of a triangle consists of the line segments and the procedures of arranging them. The construction '20–8' consists of the same elements as '8–20', including the same subtraction function. And yet, these are different constructions, for the construction rules set a different order of arguments. In the same way, by applying different construction rules to the same elements, we construct different geometrical objects.

How to understand the concept of construction as a procedure for arriving at a goal? Let us consider the case of projective geometry. It consists of a set of rules localizing objects in space. However, projective geometry does not describe some prefixed properties of some absolute space. It constructs a space. Geometric projection does not represent localizations of some points, as if there were some absolute space with fixed localizations. It represents the ways of localizing these points by determining the parameters of the space. In Riemannian terms, to determine the spatial properties of some manifold, we have to apply a geometric measure. One of such measures is projective geometry.

Generalizing this example, constructions can be described as ways of localizing the target by determining the parameters of some space. These parameters, called here construction parameters, are (a) the basic units of construction, (b) the order between (a), (c) and the relation between (a), where the basic units of construction are interpreted in the most general sense as the states of some informational space. In terms of the results of the operation, construction can be seen as the ordered triple of the construction parameters ⟨a,b,c⟩ that localize the constructed object.

> *Construction* (in terms of operation): χ is a construction of target T iff (i) χ determines the parameters of some informational space, where these parameters are (a) the basic units of χ, (b) the order between (a), and (c) the relation between (a); in order to (ii) arrive at T.
>
> *Construction* (in terms of the results of operation): an ordered triple of parameters (a), (b), and (c), consecutively.

We can localize the states of two different spaces. The logical space is the set of all possible states. Determining the parameters of the logical space is about searching for possible distinctions among ways for the world to be as well as the relations between these possible states. It is about searching for such pairs of distinctions which coincide and such which are in contradiction.

The physical space is a set of states of affairs. Determining the parameters of the physical space means searching for what the world is. Moreover, it is a search for the relations between these states of affairs.

The logical space is closely related to the physical space, for determining the parameters of the physical space is to single out one side of the distinction made in the logical space (Rayo, 2013). For instance, to construct the proposition SNOW IS WHITE, we isolate two possible states of the world – 'white-snow' and 'non-white snow' – and hold that the actual world overlaps with one side of this distinction.

The simplest case of construction is localizing a point in the physical space. Let us take a plane and try to locate a point on it. To do that, we have to determine the values of the physical space and the relation between these values that will establish the coordinate parameters. To find the localization of a point is to determine these parameters. A Cartesian diagram exemplifies this construction. It shows you how to find this localization. Interpretation of the diagram is based on the reading of the construction parameters.

Notably, a point represents spatial locations and is determined by spatial coordinates, but it is not something that is in space being projected onto the plane. Space is made of points, but it does not imply that there is some space with points in it. Space *is* a set of points. Something can be in some location in space, but localizations are not in space in the same way as chairs are. Localization does not have the property of being localized, as if there were some other localization containing it. Localizations are in relation to other localizations, in the same way as being a 5 is being in relation to other numbers, nothing less or more. To set a point in some space is to set relations of this point to others. In this sense, to determine the localizations of the points is to determine the properties of space. The Cartesian diagram does not depict some prefixed space. It determines the properties of the space to localize the point. Compare it to the way rulers work. They do not represent some prefixed distances; they localize these distances by determining the properties of the space.

Interpreting construction (in terms of the result of operation) is based on localizing the construction parameters, that is, relevant sets of information, their order and relation to other sets, to identify the searched object. Additionally, constructions consist in abstracting away from irrelevant information. For instance, the construction of a triangle identifies spatial properties, such as an orientation and the length of line segments, and sets the geometrical relation between these line segments. It determines the parameters of space to localize some spatial structure, that is, a triangle. At the same time, it abstracts away from irrelevant information, such as the colour and thickness of line segments. Interpreting the triangle diagram is based on identifying the ordered triple of the construction parameters.

By the same token, constructing 12 consists in identifying the values of the logical space, arranging them in order, and applying a relevant function, such as the subtraction function, to localize the value 12. Moreover, it abstracts away from irrelevant information, such as whether apples or fingers are subtracted.

By constructing an object, we set the parameters of some space that determine what is being represented. However, constructing objects has to be distinguished from evaluating the truth values of representation. The construction '7 + 5' localizes '12', which, in turn, can be the basis of a proposition that 12 is a sum of 7 and 5. This proposition has true values. However, the very construction is not truth-evaluable.

Compare this proposition to the operation of dividing by 0. It is not a false operation. It is an operation that fails to determine the parameters of content. Here, we do not ask whether the proposition identified by the operation of dividing by 0 is true or false. We ask what value, if any, can be identified by such an operation. Depending on whether the goal has been achieved, a construction can be successful or abortive but not true or false. We can be wrong trying to divide by 0, but we are not expressing a false proposition. We are applying an operation incorrectly.

Constructions may seem to be truth-evaluable, for they can be put into descriptive contexts. For instance, I can hold that it is true that you can construct a triangle from three straight lines. However, knowing construction procedures has to be sharply distinguished from knowing a description of a procedure. Constructing an object is a matter of exercising a skill rather than knowing a description of the object construction. For instance, applying Reidemeister moves is a skill that requires practice rather than descriptive knowledge of what Reidemeister moves are.

To stress this difference, let us compare constructions with algorithms. Informally speaking, an algorithm is a description of a procedure for finding a solution in a finite number of steps. Some constructions are algorithmizable, for algorithms can represent constructions. However, constructions cannot be reduced to algorithms.

An algorithm can be characterized as a finite set of well-defined instructions on how to move from an input to an output effectively. Moreover, according to the Church-Turing thesis, every algorithm is computable and implementable in a Turing Machine. The most common way to implement an algorithm is to express it with a recursive function. For instance, subtraction is based on the algorithm made of some values, representing the algorithm's input and the recursive subtraction function. The subtraction function is applied to calculate the output of the algorithm.

How does the concept of construction fit into this picture? Constructions cannot be functions. Functions are elements of constructions. A function is a procedure of associating the elements of a domain with the single elements of the range of the function. However, a function can operate only if the elements of the function's universe are predefined. In other words, we must determine the sets it operates on to apply a function. A construction localizes the function in a certain informational space that we consider in a given situation.

Consider the operation of dividing by 0. The condition not to divide by 0 does not follow from the content of the divide function. Dividing by 0 is not a wrong function, either. It is the same function as dividing by any other number, but the function is applied incorrectly here. Dividing is a function that calculates the number of times one number is contained within another. To know that we cannot divide by 0, we have to notice that the application of the divide function for the divider 0 leads to contradictory results. The condition not to divide by

0 was introduced to our understanding of the divide function since, without it, we would get nonsensical mathematical constructions, such as 1 = 2. In other words, the application of the divide function for the divider 0 is an abortive construction, but it is not a wrong function.

Similar observations apply to algorithms. To use an algorithm, one must construct it. For instance, it has to be decided what kind of a function has to be used and what the sets representing the input and output are. Although the construction of an algorithm is algorithmizable, it cannot be an algorithm since then we would need an algorithm for the algorithm constructing algorithms and so on.

In contrast, let us think of constructions in terms of procedural knowledge. Learning constructions is a matter of acquiring the skill of applying a procedure to localize a space state rather than knowing a description of this procedure. It is a matter of exercising procedural knowledge based on recognizing parameters of some space and not exercising propositional knowledge based on the description of this space. That is why constructions have to be exemplified rather than described.

The content of diagrams

With the concept of construction, we are now in a position to offer an approximate definition of the content of diagrams. Further on, in Chapter 6, I will use this definition to characterize the iconic content.

In general, diagrams show how to localize some objects in the logical and physical space by determining the parameters of some informational space. In particular, diagrams show what happens when certain construction procedures are applied. In short, the content of diagrams can be characterized by the construction rules of the represented object. For instance, a diagram of a triangle shows that if you take three line segments and connect them appropriately, you get a triangle. A diagram is correct if it aptly represents such procedures. It is incorrect if the represented procedures do not allow you to arrive at the goal.

Consequently, to correctly interpret a diagram is to identify the construction parameters and construction rules that determine the permissible transformations of the represented object. Grasping the content of a diagram is recognizing the parameters of construction with respect to the rules of construction to localize the searched object. For instance, to understand the knot diagram in Figure 4.1

is to identify the construction rule, which specifies that you can arrive at the unknot if you apply the first type of Reidemeister move to the knot's crossings without cutting any of them. To understand a triangle diagram is to identify the construction rule that specifies the construction parameters of the triangle. In sum, the content of diagrams consists of the construction rules used to localize a constructed object.

> *An approximate definition of the content of diagrams*: diagram *D* represents target *T*, iff *D*'s content *C* identifies (i) the construction that allows one to arrive at *T*; with respect to (ii) the rules of construction.

The content of a diagram identifies the construction rules of the object rather than denoting the object it stands for.[6] Thus, two diagrams have the same content iff they express the same construction rules instead of denoting the same objects. The terms '7 + 5' and '20–8' represent different constructions, although both denote 12. The number 12 is individuated by different constructions.

The most obvious illustrations of this definition of content are architectural drawings, manuals and maps. The standards of correctness for the architectural drawing of a building are determined by such construction rules that allow us to erect it (according to the rules of production) and identify it (according to the rules of interpretation). An incorrect architectural drawing is such that it does not allow us to construct or identify a building. Interpretation of the drawing is based on identifying the construction parameters and the construction rules that determine the permissible ways of transforming these parameters. Manuals teach us how to build something step-by-step and how to identify the relations between the elements of the mechanism. A correct manual is the one that takes us from an initial state to the goal state and helps us to identify the represented mechanism. A map is correct if it aptly represents the ways we can localize some points in some space. A correct interpretation of the map is based on identifying these ways.

To get a broader picture, consider two less-intuitive examples. The line graph displayed in Figure 4.6 represents inflation growth. It identifies the inflation rate, mapping it on time. These are the construction parameters.[7] Inflation growth is an abstract object localized by construction procedures setting the partial order between the parameters of the inflation rate and time. The graph represents inflation growth by identifying the target, which is an increasing function associating the inflation rate with time. Understanding this graph is identifying the construction parameters and the rule that if you map the inflation rate on time, the inflation rate will eventually rise. The diagram localizes the inflation

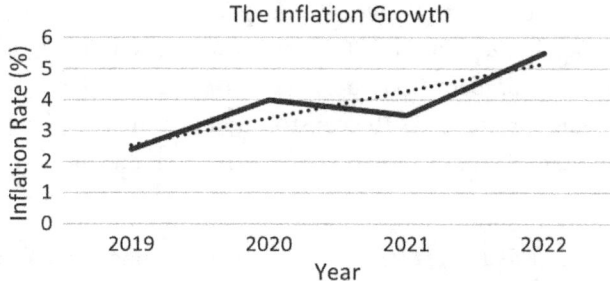

Figure 4.6 Line graph representing inflation growth. © Piotr Kozak.

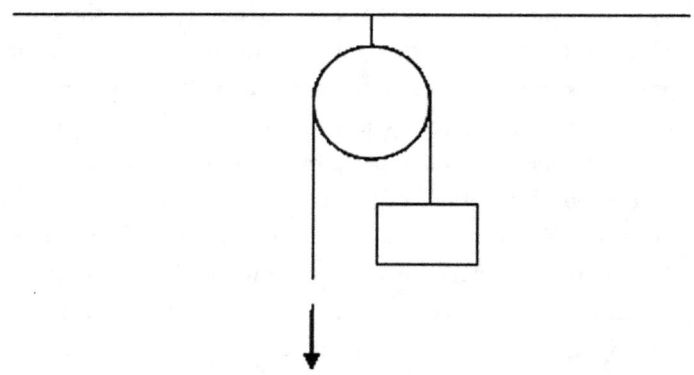

Figure 4.7 The pulley system diagram. © Piotr Kozak.

growth in relation to the inflation rate and time. Vertical and horizontal axes indicate relevant information. The irrelevant information is the colour of the line.

Next, consider the pulley system diagram represented in Figure 4.7. It shows how the mechanism behaves if the string is pulled down. The correct interpretation of the diagram is based on identifying the construction parameters: the information on the orientation and direction of the string as well as the relation between them. The visual shape of the weights is irrelevant information. Understanding this diagram consists in grasping the rule that if the string is pulled down, the weight goes up. The movement of the string can be visualized. Yet, the visualization is only a means to grasp the diagram's content. Understanding the pulley system diagram is about knowing how the system behaves if you pull the string.

Thus, the inflation and pulley system diagrams represent constructions. They show that if you identify the basic construction units and connect them according to the rules, you arrive at the target. The correct interpretation of these diagrams is based on identifying these rules.

One caveat is in order. There is a difference between Figures 4.6 and 4.7. The inflation diagram represents specific data and the relation between them. It represents a construction rule for particular values. It is equivalent to a diagram of a particular token, such as a diagram of a particular triangle. Let us call it a 'token-diagram'.

In contrast, the pulley system diagram represents a general rule of how a system behaves if the string is pulled down. It abstracts away from particular values of weight or shape. It represents a general construction rule for building a type of mechanism. It is equivalent to a diagram of a triangle as such. Let us call it a 'type-diagram'.

Every diagram can be interpreted as either a type-diagram or token-diagram.[8] For instance, an architectural drawing can show how to build a particular building, but it can be taken as showing how to build a type of building. The inflation diagram can be taken as a type-diagram if it represents the general rule of how inflation rises over time. The pulley system diagram can be taken as a token-diagram if interpreted as representing a particular mechanism.

Knowing how to construct a type-diagram does not imply constructing a token-diagram. I can know a general rule for constructing triangles without constructing any particular one in reality. Moreover, a token-diagram can never express the generality of type-diagram content.

However, every type-diagram requires the existence of a token-diagram to be represented. For example, to show how to construct triangles as such, I have to use a triangle token-diagram. Having a type-diagram implies that I have the *disposition* to construct and recognize the token-diagram of a relevant kind.[9]

Properties of constructions

A construction procedure has to be distinguished from the constructed object, for there can be constructions that construct no object. Nothing is constructed when dividing 12 by 0. Yet, it does not mean there is no construction of dividing 12 by 0. It is an abortive construction that has no denotation. In this sense, it is a different construction than '12−12', which denotes 0. An abortive construction fails to single out its referent. The construction '12−12' referent is an empty set.

This implies that constructions' properties differ from the properties of constructed objects. The '12–12' construction consists of 12 and the subtraction function. The number 0 denotes an empty set. Yet, 0 does not consist of 12 and the subtraction function. If the content of the number 0 consisted of the subtraction function and all possible values of the function, then it would have to consist of infinitely many kinds of functions that lead to 0. Instead, applying the subtraction function is the way of arriving at 0.

Consequently, knowing the content of a diagram is knowing the construction rules, not the properties of the constructed object. Knowledge of the construction rules for an object is irreducible to the knowledge of any description of that object. It covers the Euclidean idea that knowing how to construct a triangle is irreducible to knowing the definition of a triangle. Moreover, this explains why one can learn how to identify triangles without knowing what triangles are. Once they know how triangles are constructed, children can distinguish them from squares without knowing that triangles are polygons with three edges and three vertices.

The misidentification of the properties of constructions with the properties of constructed objects gives rise to the false belief that our thoughts are composed of the word- or image-like objects. Words and images are the means by which we construct thoughts. That is the idea expressed by Wittgenstein–Ryle's sceptical argument.

At the same time, this does not mean that there exist some intramental objects, such as thoughts, that can be described by words and images. Thoughts are abstract objects, such as numbers, which are the effects of construction and do not exist outside the construction context. The question 'where are thoughts' is as nonsensical as the question 'where are numbers'.

Moreover, the misidentification of these properties makes us vulnerable to falling into the pictorial fallacy. Images show how objects can be constructed and how they would look if they were constructed in some way. Yet, this does not imply that they describe these objects.

Investigating the form of constructions is cognitively productive. Manipulating the form of constructions can lead to the discovery of their novel properties. Seeing that $a + b = b + a$ cannot be inferred from the content of the symbols. Discovering the commutative property of addition entails the capacity to recognize that changing the order of the elements does not affect the result of the construction.

At the same time, constructions are characterized as operations determining the parameters of some informational space. Thus, different constructions can highlight different aspects of information, enriching our understanding of constructed objects.

Let us consider a geometrical example. Note that the Euclidean condition that any two line segments of the triangle together are longer than the third is not analytically derivable from the definition of a triangle. The content of the definition – 'a polygon with three edges and three vertices' – does not contain the postulate that any two line segments together are longer than the third. Euclid introduces this postulate, for, without it, the construction of the triangle is impossible. In other words, we know that this postulate is part of the content of the triangle concept because, without this postulate, it would be impossible to construct a triangle.

Thus, investigating the way in which mathematical objects can be constructed teaches us something about their nature. That is why diagrams can be useful in the context of discovery in mathematical practice. Particularly, interpreting the content of diagrams in terms of the rules of construction helps us to clarify the concept of aspect shifting in diagrams. Note that diagrams can be rearranged, which makes them open to new interpretations. The same diagram can be arranged differently, which allows for addressing its various properties. Giaquinto (2007) recognizes this as a crucial aspect of visual thinking in mathematics. For Peirce, this is an iconic property of diagrammatic reasoning (*CP* 2.279). For instance, manipulating the construction of triangles helps to find a proof strategy for Pythagoras' Theorem.[10]

However, it is not clear what these different aspects of mathematical objects are. According to the view presented here, they can be identified by means of different constructions. Different interpretations of a diagram single out different construction parameters.

Let us consider the well-known Necker cube, an ambiguous figure with two different interpretations. We can understand the content of these two interpretations in terms of displaying different constructions. We can pick out the lower-right and upper-left faces and arrange them according to the 'being-in-front' relation. We get two different constructions depending on the order of the arguments in this relation. In the same way, we can have two different constructions for the subtraction function and arguments '20' and '8'.

Thus, different aspects of information can be identified by using different constructions. For instance, line and bar graphs (Figure 4.8) use the same data. Yet, they are differently constructed, affecting the interpretation of the information.

The line graph (a) makes it easier to identify the relationship between exercise and weight loss. An increasing function connects the data. In contrast, bar graph (b) makes it easier to single out the relationship between exercise and the

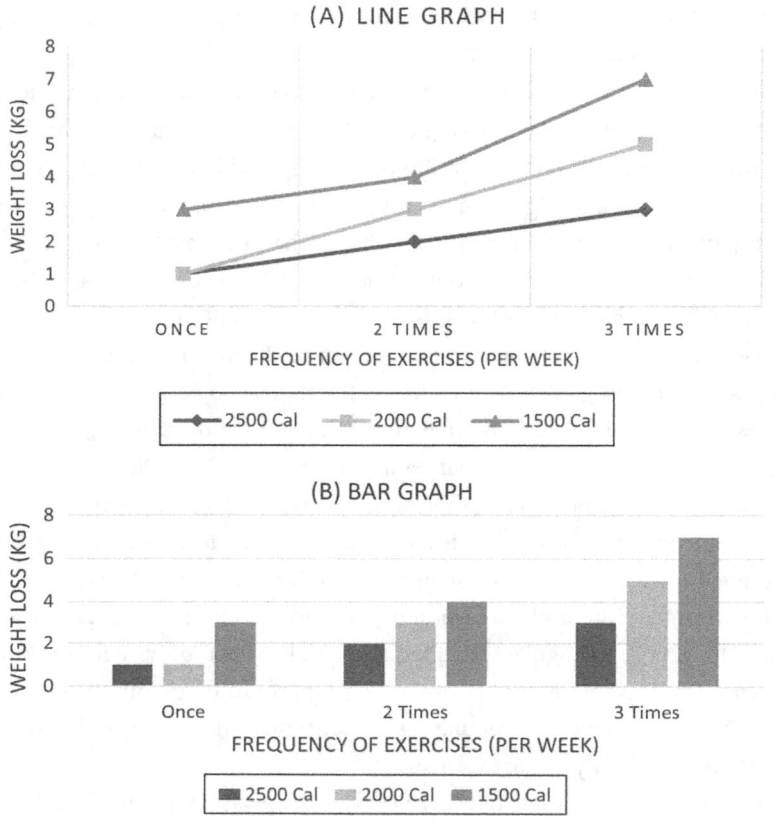

Figure 4.8 Line graphs (a) and bar charts (b) convey the same information but in a different way, affecting the accessibility of the information. © Piotr Kozak.

number of calories burned because the bars comparing the data for calories and the amount of exercise are closer to one another. The line and bar graphs employ the same data but construct it differently. The graphs highlight different aspects of information, depending on the applied construction.

Explaining resemblance

In his *Syllabus* (*CP* 2.277), Peirce defines diagrams as the representations that refer to relations via resembling them. Although this characteristic is, in principle, not false, it is a source of a deep misunderstanding.[11]

It is far from obvious what we mean when we say that diagrams resemble anything. The resemblance is usually understood as sharing first-order properties of objects. The diagram representing inflation growth does not resemble inflation growth. Inflation growth is an abstract object that cannot be a relatum in the relation of resemblance.

At first glance, we can try to explicate the concept of resemblance in terms of structural similarity. The concept of structural similarity denotes sharing the same structure in the following sense. Two sets, A and B, are structurally similar iff there is a function mapping (some) elements of A and (some) relations defined over the members of A onto (some) elements of B and (some) relations defined over the members of B such that preserves the second-order properties of the relations in A. Thus, if some relation holds for (some) elements of A, the corresponding relation holds for (some) elements of B (e.g. Kulvicki, 2014; Swoyer, 1991).[12] Consequently, two objects, a and b, are structurally similar if and only if various relationships among the parts of a correspond to important relationships among the parts of b.

Structural similarity is not a sufficient condition to determine the content of a diagram. A trace left by an ant can be structurally similar to Winston Churchill, but it does not refer to Winston Churchill (Putnam, 1981). Yet, structural similarity seems to be a necessary condition of being a diagram and having content.

To illustrate the idea of structural similarity, let us consider Euler diagrams. The core idea is that we can represent subset, intersection and disjoint relations between sets in terms of inclusion, overlapping and disjointness of circles (Figure 4.9). It may seem that the concept of resemblance is crucial to explain why we can represent abstract relations by Euler diagrams. According to this view, Euler diagrams represent logical relations because they mirror the logical structure of these relations.

However, the concept of structural similarity is not necessary to explain how Euler diagrams work. We can explain it by pointing out that they represent the same construction rules. For instance, we can represent the subset relation in terms of the inclusion of circles as they represent the same constructions.

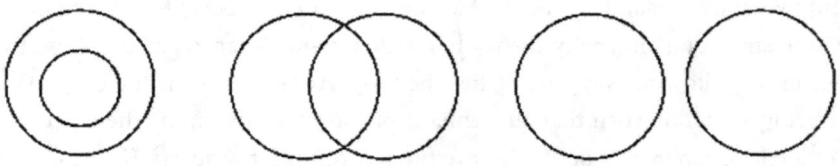

Figure 4.9 Euler diagrams representing the subset, intersection and disjoint relations, respectively. © Piotr Kozak.

The subset relation is constructed according to the same rules as the inclusion of circles in the following sense. You can arrive at subset A if you include all elements of subset A into set B. In the same way, you can arrive at the inclusion of circles if all elements of one circle are contained within another.

Explaining how Euler diagrams work in terms of sharing the same construction rules has a significant advantage over explaining diagrams in terms of resemblance. It allows us to avoid the following categorical fallacy. Note that if one uses the resemblance relation to explain the fact that the subset relation can be represented in Euler diagrams by the inclusion of circles, one holds that the subset relation and the inclusion of circles are identical in some respect.

However, that is nonsense. Subsets do not share any non-trivial properties with the circles represented in Euler diagrams. Subsets and drawings belong to different ontological categories. Subsets are logical, while drawings are physical objects. Claiming that they are alike is as much nonsense as claiming that numbers share some properties with apples. Apples can be red, numbers can be even, but not the other way round. Numbers can represent the quantity of apples, but it does not imply that they are alike. By drawing circles, we can represent subsets, but it does not imply that the representation and the represented object share some common properties.

Seemingly, the categorical fallacy objection does not apply to the structural similarity theory. After all, it does not hold that logical objects, like relations and numbers, and physical objects, like drawings and apples, share some first-order properties, like being red or tasty. The idea is that these objects share some second-order properties. Thus, a subset is structurally similar to a drawing of two circles as some second-order properties, such as being an asymmetric relation, are preserved. Second-order properties do not characterize the objects of first-order properties. An apple can be red (first-order property), red is a colour (second-order property), but an apple is not a colour. Therefore, the ontological fallacy argument is misplaced.

Still, this gives us an unattractive ontology of mathematical objects. Compare two sets, A and B, where B has the first-order property F of being a subset of A, and two objects, a and b, where b has the first-order property G of being inside a. The structural similarity theory holds that a and b can represent these two sets in the following way. First, b has the property G, and G has the property H of being a relation such that all elements of b are included in a. The object b is not a relation, hence, it does not have the second-order property H. Second, set B has the property F, and F has the property H of being a relation such that all elements of B are included in A.

This story suggests that there is some logical object, 'a subset', that can be described by the relation of being included. This does not seem right. The term 'being included' is only another way of saying that something is a subset. The inclusion relation is not the property of being a subset as if there were some logical objects described by this property. The inclusion relation is a way of isolating some elements of a certain kind in a set. The effect of such a procedure is a subset. But a subset is nothing more than 'being included'. Something cannot be a subset without being included.

In contrast, b is an object that can be identified by the property of being inside a. However, b is not equal to the property G of being inside a. We use the property G to identify b. Yet, the object and its property are distinct. The object b could have no property G.

What is the moral of that? Note that the story told by the structural similarity theory is plausible only if we adopt some sort of mathematical Platonism regarding the nature of mathematical objects. It does not have to be wrong. It is not necessarily correct either. However, it would be unfair to insist on adopting Platonism only for the purposes of explaining how Euler diagrams work.

These ontological problems are avoided if we explicate Euler diagrams in terms of representing the same construction rules, for the properties of constructions are not the properties of the constructed objects. Euler diagrams and logical relations can share the properties of constructions without sharing the properties of objects. We do not have to be Platonists, which is fair.

Moreover, resemblance-based theories give us the wrong idea of the epistemology of diagrams. First, it is not clear how resemblance-based accounts can distinguish between misrepresentation and non-representation (Suárez, 2003). To misrepresent something is to assign the represented object with properties that it does not have. A non-representation does not assign any property. According to resemblance-based accounts, representing something is sharing first- or second-order properties. However, if something does not share any property with the represented object, it is not a representation. This seems to be clearly wrong since misrepresenting a circle with the diagram of a square does not imply that a diagram of a square is not a representation.

We cannot simply bypass this problem by invoking the idea of preserving a partial structure (e.g. Bueno and French, 2011), for this idea is too inclusive. For one thing, there is a partial structure-preserving relation between a representation of a dot and a diagram of a circle and a square. Yet, it does not mean that the single-dot diagram represents circles and squares. For another, there is no way to distinguish between partial representation and non-representation.

A scribble does not represent anything, yet it preserves some partial structure with any object.

Second, the concepts of resemblance and structural similarity are too weak to identify the represented objects. In order to know what is being resembled, we have to identify relevant aspects of resemblance. Yet, according to the well-known Wollheim's argument (1998, 2003), to identify relevant aspects of resemblance, one has to identify what is being resembled. In Wollheimian terms, recognizing an object depicted on a picture's surface presupposes that we have a seeing-in experience of this object. Provided that we have such experience, we can recognize the depicted object in the marks on the picture's surface. If seeing-in were not prior to the experienced resemblance, there would be no way to discern accidental and non-accidental marks on the surface.

Consider an ellipse-shaped diagram. It is structurally similar to a circle seen from a perspective. Without knowing what this diagram represents, we cannot know which parts of the structure are relevant. However, it was by comparing structures that we were supposed to identify the object of reference.

At first glance, this should not be particularly disturbing. Most of the proponents of resemblance-based theories acknowledge that sharing properties is not a sufficient condition of representation. It is held (e.g. Abell, 2009; Blumson, 2014; Bueno and French, 2011; van Fraassen, 2008) that the epistemological problems can be resolved by supplementing the condition of sharing properties with the pragmatical condition that the representation has to be used by an agent in a proper way. The general idea is that to recognize the relevant aspect of resemblance, one has to recognize the intention of the agent.

However, such a move only replaces one problem with another. Note that we shift the burden of the argument from resemblance-based accounts to intention-based accounts, which have their own problems, and, as Wittgenstein's argument from content indeterminacy shows, cannot explain depiction.

Moreover, such a shift renders the resemblance relation functionally irrelevant. The resemblance does nothing in the process of recognizing the sign's content, for all we have to do is to recognize the agent's intentions. Without any supplementation, the intention-condition is insufficient.

Explaining the content of diagrams in terms of the rules of construction not only gives us a better picture of the content. It avoids the epistemological problems of resemblance-based accounts and provides us with an explanation of what resemblance is. Most of all, it allows us to make sense of the concept of resemblance without sharing first-order or structural properties.

Let me illustrate this with the following example. Take two diagrams represented in Figure 4.10. The dashed line represents unfinished lines. Note that they lack similarity in shape and structure since some lines are missing. However, they represent the same construction rules, thus having the same content. For it is possible to construct a triangle if the straight lines are drawn further in thought. The fact that we run out of space on paper plays no role in constructing triangles.

In contrast, two identical ellipse-shaped diagrams can denote two different objects: either an ellipse or a circle seen from a perspective. These diagrams have the same spatial properties but different content, for different construction rules are applied.

Next, depending on the kind of identified constructions, we can determine the relevant aspects of resemblance. Consider two triangles $\triangle ABD$ and $\triangle ACE$ in Figure 4.11. Do they resemble each other? It depends on the rules of construction we recognize as relevant. Suppose we want to know whether they are geometrically similar. In that case, we are interested in the construction rule, which identifies as relevant the information on the ratio of the sides' lengths and the measures of the angles. If we want to know whether they are congruent,

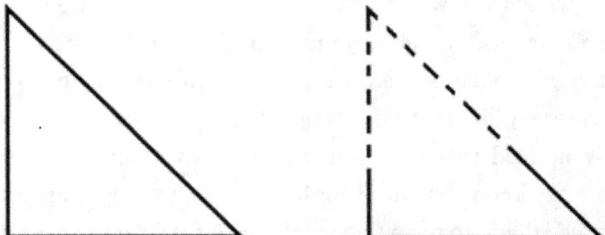

Figure 4.10 Two diagrams of a triangle with the same construction rules. © Piotr Kozak.

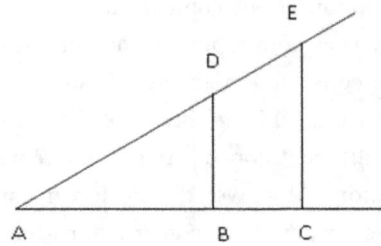

Figure 4.11 Depending on the identified rules of construction, $\triangle ABD$ and $\triangle ACE$ are similar or congruent to each other. © Piotr Kozak.

the construction rules single out the information on shape and size. $\triangle ABD$ and $\triangle ACE$ are similar or congruent depending on the identified construction rules. Consequently, the diagrams $\triangle ABD$ and $\triangle ACE$ represent the same or different objects. In the most general sense, any two token-diagrams of triangles are similar if they depict the construction rules for any polygon with three edges and three vertices.

Finally, the idea of construction gives us an insight into Gombrich's so-called riddle of style. According to Gombrich (1960), adopting the resemblance-based view of depiction does not allow us to address the question of why people at different times and in different cultures tend to depict objects in different ways. Gombrich argues that this riddle is based on a false assumption that depiction is based on capturing a visual likeness of the depicted objects. Based on this assumption, pictorial representational systems should have developed in one direction, perfecting the ability to express the resemblance so that eventually, there should be one style that unifies all others and represents the world most adequately.

To address the riddle of style, Gombrich acknowledges that depiction is based on recognizing familiar objects in patterns of lines and colours, which elicits an illusion of seeing these objects. To recognize objects in these patterns, we have to employ schemata, that is, stereotypic ways of depicting objects. Schemata are a means to represent the world in the same way as projection in cartography serves to represent certain regions in the world. Different schemata are used to represent different aspects of the world, just like different methods of projection are used to represent different aspects of space.

It is rarely noticed that the same riddle of style applies to diagrams.[13] If diagrams were based on structural similarity, then we might expect a tendency to create a unified diagram form that would preserve structures most adequately. That contradicts the common observation that there is a reverse trend in creating many different forms of diagrams.

The concept of construction covers Gombrich's intuitions without applying the vague illusion-metaphor. Let us compare the line graph and the bar chart displayed in Figure 4.8. They both represent the same structure of information. Yet, depending on their construction, the aspects of information they represent are different. We expect that different styles of diagrammatic representations will coincide with the different constructions they represent, which seems to agree with our observations. Moreover, the illusion-metaphor can be explicated in terms of sharing the properties of constructions but not the properties of objects. That seems to align with Wittgenstein's and Frege-Davidson's argument of lack of truth values.

Where do we stand with the concept of resemblance? Much depends on how we understand resemblance. It is an umbrella term with shifting reference being insensitive to counterexamples. It can be understood in relational and non-relational terms (e.g. Hyman, 2012), as a symmetrical and asymmetrical relation and so forth.

However, saying that resemblance cannot be an explanation of how images work is not saying that we can dismiss resemblance as an insufficient condition of how images represent; or that images do not share properties with the represented object. I hold that images do share a structure with the represented objects.

Moreover, I hold that resemblance plays an essential role in the process of iconic representation, and the fact that an image resembles the represented object has to be explained by any full-fledged theory of depiction. If our explanation of depiction were unable to explain the concept of resemblance, it would be considered insufficient; for the concept of resemblance is the intuitive starting point of any theory of depiction (Schier, 1986; Walton, 1973). It is enough to say that the inability to explain how images can resemble their objects is the chief reason why the Goodmanian theory of depiction seems to fail.

The theory of construction can explain how diagrams resemble represented objects. The resemblance between any two objects can be explained in terms of representing the same construction rules. Although the concept of resemblance cannot explain how diagrams work, the concept of construction gives us an idea of how diagrams can resemble their objects and have content. In a nutshell, the resemblance relation does not determine the construction rules. On the contrary, the construction rules determine the resemblance relation. Diagrams resemble the relations in terms of representing the same construction rules. To determine what is being resembled, one has to determine what the rules of construction are.

Summary

Based on the analyses of the content of diagrams, I demonstrate that diagrams are necessary for representing the construction operation. I explicate the concept of construction in terms of procedures of arriving at a goal by means of determining the parameters of some informational space. Accordingly, I distinguish between the properties of constructions and the properties of objects.

I describe the relationship between the concept of construction and resemblance. I claim that the presented description of content is able to explain the concept of resemblance without taking this concept as an explanans of the content of diagrams.

5

Recognition-based identification

In the previous chapter, based on the analysis of diagrams, I pointed out that images exemplify the ways of constructing objects. In this chapter, I discuss the second role of images: triggering the ability to recognize objects and events. Drawing on Evans, I introduce the concept of recognition-based identification. I argue that images can convey a reference relation distinct from demonstratives and descriptions. It can be explicated based on the construction concept. I hold that recognition is based on identifying the construction invariants which determine the properties of some space in order to localize the depicted object.

Case Study 2: The picture of a black hole

When the Event Horizon Telescope (EHT) caught its first glimpse of the M87 black hole, located over fifty-four million light years away from Earth, one could almost hear a moan of disappointment. After all, did we not have more detailed pictures before? Simulations of black holes had looked more stunning. Is the blurry picture of a hazy cosmic flame loop all that we can get?

The picture represented in Figure 5.1 may not seem like much, but it is a breakthrough in physics. Why does having this picture matter?

The black hole picture helps us validate theories and measurement methods. Up-to-date simulations of black holes based on Einstein's theory of general relativity have been commonly used in science to predict how a black hole behaves. As Einstein's theory predicted, the shape of a black hole would be almost a perfect circle, and its size should be directly related to its mass. As it turned out, it is. The picture has allowed us, for the first time, to test Einstein's theory of relativity for black holes and other regions containing dense matter.

Furthermore, the picture is an object of investigation in that we now have a way to study black holes as physical objects rather than mathematical constructs.

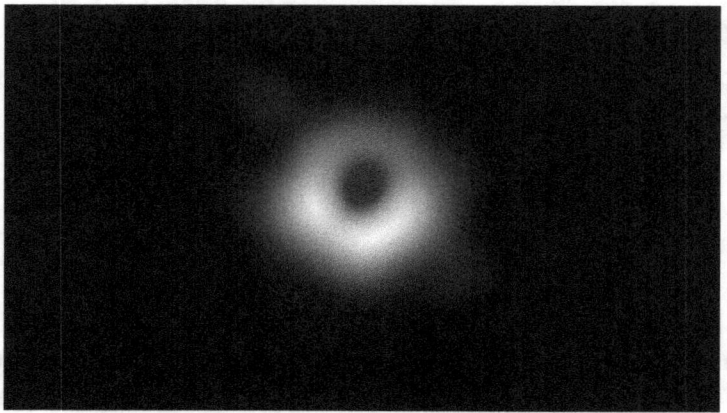

Figure 5.1 The first picture of a black hole. Credit: Event Horizon Telescope collaboration et al. National Science Foundation.Source: European Southern Observatory (ESO) via Wikimedia Commons.

Thanks to the picture, we are able to calculate the black hole's mass more precisely. Using the EHT, we are able to measure the radius of M87's event horizon directly and calculate the hole's mass. Consequently, the calculation helps us validate available methods of mass estimation.

Additionally, the picture shows us the features that equations have not yet captured. For example, there is the open question of why the black hole's silhouette is irregular in shape. It is a question that we hope to answer in the future. It is also believed that the data gathered by the EHT will offer some insight into the formation and behaviour of black holes. The structure revealed in the image can help to reveal how material gains the energy needed to fall into a black hole or spiral away in jets. Lastly, it is believed that information provided by the EHT will help us resolve fundamental physics problems, such as the black hole information paradox.

Thus, the black hole picture is much more than a blurry illustration of a night sky. Let us think of it as the final result of a measurement operation. The picture represents measured features of the object, where the measurement values can be projected onto the measured object. We study the features of the picture to learn something about the features of the black hole. Before the picture was taken, some facts were unknown and could not be deduced. If they could be deduced, physics would be reducible to metaphysics (e.g. Elgin, 1996a, 2017).

However, the fact that the picture may seem underwhelming to some teaches us something important about our understanding of what images are. Let me illustrate it with Dennett's (1990, 1991a) argument from computer reasoning.

The black hole had been first measured and described in the numerical system of representation, which, in turn, was presented as the black hole picture. Displaying the information on the picture does not change the content of the information. Before we displayed the information on the properties of the black hole, we had had the information.

Moreover, based on the syntactic properties of numerical data, we can compute the non-measured properties, such as the mass of the black hole. Displaying the information via the picture does not affect the content. Therefore, having the black hole picture is inessential for knowledge and reasoning. To show what is wrong with this argument, let us turn to the idea of the naturalness of icons.

Naturalness of icons

The backbone of Dennett's argument is the belief that depictions and numerical representations carry the same kind of content since both code the same information. The only difference is the way the information is displayed. This belief is based on Goodman's idea (1976) that images and language-like representations are different kinds of a conventional representational system.

However, this does not seem right, for it fails to explain the properties of the content of images. In contrast to language-like representations, images seem to bear some natural relation to the represented objects, which explains why a depiction of a horse looks like a horse. The language-like representation of a horse does not have to be like a horse.[1] What does it mean to bear a natural relation?

The notion of 'natural relation' is deeply ambiguous. In the sense that interests us here, it refers to a nonconventional relation. There are, however, two understandings of the convention concept. The first holds that a sign is conventional iff it has arbitrary semantics. The second holds that a sign is conventional if it belongs to some representational system. It is noncontroversial that images are conventional in the second sense, for they can be easily classified according to different representational styles (Willats, 1997). It is highly controversial whether they are conventional in the first sense. The concept of a natural relation will be understood here as bearing a non-arbitrary relation to the represented object.

The nonconventional relation between a sign and the represented object can be explicated in two ways. First, it can be put in psychological terms (e.g. Burge, 2018; Giardino and Greenberg, 2015). According to the psychological understanding of this term, a representational system is more or less natural to the degree to which natural psychological abilities, rather than cultural factors, make the system easy to use. For instance, it is easier to represent directions with left and right arrows instead of letters A and B, for the arrows better fit our psychological abilities.

However, the psychological understanding is not of the appropriate modal strength to explicate the idea of a non-arbitrary relation. It is metaphysically contingent matter what kind of psychological abilities we possess. If we were robots, it could be more practical to internalize letters instead of arrows. Such defined naturalness is only the complement of the metaphysical arbitrariness of the representational systems.

Second, naturalness can be understood in metaphysical terms. The representational system is natural only if the properties of the system depend on the properties of the represented object. The psychological naturalness of icons can be a simple consequence of the metaphysical naturalness so defined. What does it mean for the representational system to depend on the represented object? Let me explain it with the help of the concept of natural generativity.

According to the recognition-based theories of depiction (e.g. Lopes, 1996, 2005; Sartwell, 1991; Schier, 1986; Squires, 1969), the identification of depicted objects is generated naturally. The concept of natural generativity refers to the observation that if someone has interpreted any member of an iconic system of representation, then one can interpret any other member of the system, provided that one can recognize it in perception. As Wollheim puts it (1987, 77): 'if I can recognize a picture of a cat, and I know what a dog looks like, then I can be expected to recognize a picture of a dog.' Knowing the meaning of one part of an iconic system implies knowing how to interpret the meaning of the other parts.

The idea of natural generativity marks the difference between iconic and symbolic systems. Symbolic systems lack natural generativity. They are based on arbitrary semantical and syntactic rules that have to be taught one by one. Thus, the knowledge of a symbolic system cannot be derived from the knowledge of a particular part of the system. Knowing that German 'Hund' means a dog and how cats look does not imply knowing what 'Katze' means. I do not understand what 'Katze' means if I did not encounter this word before. In contrast, I can understand pictures I have never seen before.

According to Schier (1986), any theory of depiction should be able to explain the natural generativity of pictures. The recognition-based theories can provide such an explanation granted that recognitional capacities triggered in picture-perception are the same as in the case of perception in the flesh. Pictures cause naturally generated interpretations by activating the recognitional skills of the competent spectators. In short, picture P depicts an object O only if P elicits competent spectators' capacities to recognize O. We can understand pictures for the ability to understand them piggybacks on the everyday recognitional skills, exploiting the resources normally deployed to identify objects in perception (Lopes, 1996, 2003; Schier, 1986).

What can we learn from recognitional theories of depiction? Drawing on Schier, I hold that the irreducible role of images is to trigger the recognitional capacities of the interpreter. A successful interpretation of a picture involves the capacity to recognize the depicted object. We see objects through their depictions (e.g. Aasen, 2016; Lopes, 1996, 2005; Walton, 1984; Wollheim, 1987). Symbols cannot do that. Images convey a special kind of reference relation which is not accessible for symbolical representations. Thus, images are necessary for the operation of recognition. The fundamental question of any theory of depiction is how to explain this reference function (Budd, 2008; Newall, 2011).

Let us go back to the black hole picture. Unlike numerical data, the picture makes it possible to refer numbers to physical properties and recognize them in the physical space. The picture offers us a direct link to the investigated object, mapping the logical space of numerical data onto the physical space. The black hole picture triggers recognition abilities and enables us to relate abstract models to physical reality.

However, that cannot be the whole story. First, suppose we follow Schier and interpret recognition in psychological terms as a set of perceptual abilities. In that case, the recognition-based theory is useless when explaining how we understand the black hole picture. We cannot see a black hole in the flesh. The black hole picture cannot trigger the same perceptual skills as in the case of perception.

Second, various theories of depiction are compatible with the claim that pictures activate the same recognitional capacities as in the case of perceiving the depicted object in the flesh. The question is how pictures can do so.

The most intuitive alternative to recognition-based theories are resemblance-based theories. According to resemblance-based theories, the most natural explanation of the fact that pictures trigger recognitional capacities is that pictures resemble the depicted objects, where the resemblance is usually understood as sharing first- or second-order properties with the represented object (Hopkins, 1998; Hyman, 2006; Neander, 1987; Newall, 2011; Novitz, 1988). However, resemblance-based theories cannot explain how we can recognize a black hole. Determining the resemblance relation presupposes the ability to make a comparison between the relata. In the case of the black hole picture, such a comparison is impossible. Pictures are commonly used to depict unobservable phenomena (Schöttler, 2017). Resemblance-based theories cannot explain how this is possible. Thus, we need a better (and non-psychological) explanation of recognition.

What is recognition?

The concept of recognition-based identification was introduced by Evans (1982) to distinguish this kind of reference from demonstratives and descriptions. Recognition is an operation that makes it possible to reidentify an object or event. To explicate the recognition concept, let me start with some distinctions.

Let us consider seeing the black hole in the picture. We can distinguish between three types of perception here. First, we can *perceive* the black hole. Second, we can *recognize* the black hole. Third, we can see *that* it is a black hole.[2] Drawing on Hope (2009), I call these object perception, perceptual recognition and that-perception.

All three types of perception are functionally distinct (Overgaard, 2022). I can see a black hole, but I do not have to see that it is a black hole. I can mistake it for a picture of a donut. Seeing a black hole does not entail recognizing a black hole. I can see something without recognizing it. Recognizing a black hole does not entail seeing that it is a black hole. I can recognize the picture of a black hole as similar to other pictures without seeing that it is a picture of a black hole.

Object perception is basic to others. I cannot recognize a black hole and see that it is a black hole without perceiving the black hole. Object perception can be described as the ability to fix perception on some objects and scenarios.

Object perception consists in extracting a perceptual object from the background and distinguishing it from other perceptual objects, a process that underlies perceptual selection and tracking of objects (e.g. Casati, 2015). In order to do that, the representational system has to identify the perceived object demonstratively. For instance, it has to be determined that the texture of a picture's surface is irrelevant to object perception. Next, the system has to determine topological relations to set part–whole relations between perceptual objects. For one thing, this consists in distinguishing the whole object from the background and the other objects; for another, in identifying the parts of the whole object.

Misrepresentation of object perception consists in the inability to identify the object. For instance, I may not be able to distinguish between the features of the depicted object and the features of the picture's surface or between the object and the background.

Perceptual recognition describes the ability to associate perceptual objects with each other and reidentify an object after some changes. It is often described as recollecting information and having a sense of familiarity with the recognized object (e.g. Dokic, 2010; Hope, 2009; Hummel, 2013). Here it is understood as a perceiving that perceived objects belong to the same kind.[3]

Misrecognition can be easily confused with errors in that-perception. For instance, looking at the black hole picture, I can misinterpret it as a picture of a donut. Yet, errors in perceptual recognition are distinct from errors in that-perception. The inability to recognize two objects as belonging to the same kind is different from the inability to identify the kind they belong to.

Perceptual recognition is a non-propositional, non-conceptual and non-linguistic form of perception. A child can recognize triangles without knowing that they are polygons with three edges and three vertices or that they are called 'triangles'. In order to do that, the child must learn that some spatial properties remain the same across changes in a different context.

This ability is non-propositional. Children can be wrong in confusing triangles with squares, but they are not expressing any false proposition. It does not require concepts (e.g. Siewert, 1998). It is, rather, a prerequisite for having concepts (Peacocke, 1992; Giovannelli, 2001). Before one can subsume an object under a concept, one has to be able to recognize this object. Recognition is a gateway from object perception to cognitive processes such as categorization and reasoning, but it is not based on these processes.

Only that-perception requires the presence of language-like, conceptual and propositional structures (e.g. Armstrong, 1997). If I see that it is a black picture, it must be possible to express it in the language. It is conceptual since it involves subsuming a representation under a concept. It is propositional since perceiving states of affairs is truth-evaluable. If I am wrong when I see that it is a black hole, then it is not true that it is a black hole.

Object perception, perceptual recognition and that-perception of the black hole use the same kind of perception. Seeing-that is based on sensory perception, but it is not another kind of perception. It is not 'perceiving the truth of propositions' as if there were a perception of propositions. Propositions are ways of identifying how the world can be. These are not the objects we can perceive. Seeing-that is, rather, a form of propositional attitude based on perception. Accordingly, I cannot see that it is a black hole without having a belief that it is a black hole. Seeing that it is a black hole and believing that it is not a black hole is self-contradictory.

In contrast, I can recognize a black hole without having a belief that it is a black hole. I can recognize my mother in a medieval painting without believing that it is my mother. However, in accord with Russell's Principle, I cannot have that-perception without the recognition of the object of perception. I cannot see that it is snowing without the ability to recognize snow (e.g. Evans, 1982; Peacocke, 1992).

Varieties of reference

According to the influential account in cognitive science, perceptual recognition is connected with indexing an object with an object-file (e.g. Green and Quilty-Dunn, 2021; Murez and Recanati, 2016; Pylyshyn, 2007; Recanati, 2012). An object-file is a representation that sustains reference to an external object over time. It helps to explain how minds can keep track of objects in different contexts. This ability is illustrated by the well-known 'tunnel effect' (Burke, 1952), according to which if an object moves beyond a visual obstacle and reappears on the other side, it is perceived as a single entity. This is true even if the reappearing object has different physical properties from the disappearing one. Thus, there are strong reasons to believe that the information stored in the object-files is dissociated from the information on the object's features. Subjects can reliably track objects despite significant changes in colour, shape or size, which implies that object-files have directly referential (nondescriptive) semantics.

The object-file framework seems to offer an elegant explanation of how recognition works. The story goes as follows (e.g. Recanati, 2012, 85 ff.): during the first perceptual encounter with a target, we form a demonstrative file that indicates the target. Next, the demonstrative file is converted into a memory demonstrative which stores information on past encounters with the target. Recognition is based on finding an index that converts the memory demonstrative into the recognitional demonstrative, which identifies the link between the perceptual and memory information. Recognition consists in being reacquainted with a target.

This story is simple but uninformative without explaining how memory search works. First, the cognitive system cannot automatically identify the memory file, for it would have to search through all the possible files, which would soon lead to the combinatory explosion. The search cannot be based on looking for similarities between two demonstratives, as in the case of Gauker's (2011) idea of perceptual similarity spaces, for it would require the object-file to be descriptive.

Such files would have to contain information on the features of the target, which, in turn, would make them unable to explain why the ability to track objects is disassociated from the information on the object's features. Moreover, it does not solve the combinatory explosion problem without any explanation of how looking for similarities works. Interpreting the idea of similarity as primitive and unanalysable does not offer an explanation but excuses.

Second, the proponents of the object-file framework can interpret recognition in terms of identifying probabilistically determined causal links in the following way. If we systematically encounter target T producing demonstrative file $D(t)$, there is an increasing probability that it would result in producing associated memory file $M(t)$. Recognition is a matter of causally driven activation of $D(t)$ and associated $M(t)$.

However, that won't do. For one thing, the causal story is vulnerable to counterexamples. I can recognize objects I did not encounter in the past, for example, I can recognize my mother in her childhood photograph. I can recognize objects I have a wrong idea of, too. If it turned out that previous visualizations of a black hole were wrong, and the depicted black hole picture resembled a square rather than a circle, scientists would still identify it as a black hole picture.

Moreover, the story is empirically implausible, for recognition abilities seem to be functionally dissociated from the ability to retrieve information from memory causally. I can remember perceptual objects without recognizing them. For instance, patients with associative visual agnosia can successfully copy drawings but cannot recognize the copied objects (e.g. McCarthy and Warrington, 1986). Next, I can recognize a familiar scenario connected to a false memory of myself as a child being lost in a supermarket. False memories are not causally connected with any events.[4]

Thus, although the object-files framework can work well to explain the mechanisms of demonstrative reference, it does not seem to explain the nature of recognition-based identification. This suggests that recognition is not a kind of demonstrative reference.

It seems natural to think that recognition involves perception. Consequently, it may seem that recognition-based identification is a kind of demonstrative reference (e.g. Strawson, 1963). Demonstratives are based on a causal information link between a representation and its target and the ability to locate and track this target in space. This implies that to understand the content of a demonstrative, one has to identify a relevant causal link and find the target within some egocentric frame of reference, determining the here-now relation.

Indicating a target means determining the spatiotemporal relation between myself and the indicated object (e.g. Evans, 1982; Kaplan, 1989; Matthen, 2005).

Recognition-based identification cannot be a kind of demonstrative (Bermúdez, 2000; Curry, 1995; Zeimbekis, 2010). Recognition-based identification does not put anything within the egocentric space. The target of the black hole picture is not 'there' as if indicating the picture would localize the black hole. The picture's target is not located within the here-now framework, implying that we can recognize it without having a demonstrative of the target. For instance, we can recognize Santa Clause in a picture without being able to refer to Santa Claus demonstratively.

The source of the confusion is that we can demonstratively refer to the representation of the target. For example, I can indicate my mother's photograph and hold 'that is my mother'. Yet, it is not demonstrative of a target. If I place a picture of my mother next to her, I am not holding that my mother is in two locations at once. I refer to my mother by recognizing her in the picture.

However, recognition-based identification is akin to demonstratives in at least one respect. Recognition seems to differ from identification by description, which holds that we identify objects by matching them to their descriptions. However, the descriptive theory of recognition seems empirically and conceptually implausible.

It is empirically unsupported, for it cannot explain how we can track objects that change their features across temporal and spatial contexts. Notably, it faces well-known objections against template and feature theories of pattern recognition in cognitive psychology (e.g. Liu et al., 1995). The template theory holds that we recognize perceived objects by matching them to a template stored in long-term memory. However, it does not explain how we can store the information about recognized objects. We perceive objects constantly changing, which would require storing infinitely many templates of perceived objects. The feature theory holds that we recognize objects by matching the selected features of objects. However, it cannot explain how we can recognize different interpretations of bistable objects which share the same features. Moreover, object recognition skill, called 'o' skill in cognitive psychology, is dissociated from general intelligence and propositional knowledge. Neither IQ nor SAT scores can predict the recognition of objects (Richler et al., 2017).

The descriptive theory is conceptually incoherent, for I can recognize an object without knowing its description. For instance, I can recognize my mother in her childhood photographs without knowing what she looked like. Moreover, I can know a description of the object without any ability to recognize it. For instance, I can know what my aunt looks like but cannot recognize her in a photograph. The recognition-based identification seems to refer to the target directly. Recognizing my aunt in a photograph differs from knowing a description 'the woman represented in the photograph is my aunt' (Lopes, 2010; Terrone, 2021a). The question is what such recognition-based identification involves.

Construction invariants

The core idea of recognition is identifying a target by fitting it into a pattern. This pattern does not have to be recollected. It can be discovered. Recognizing a pattern is finding an invariant. It is seeing that a target remains the same across changes and contexts, including changes in viewpoint and target properties. For instance, I can recognize my mother in her childhood picture if I identify some transformation invariants, that is, the properties that remain unchanged after transformations. Depending on the context, this invariant can be a birthmark, a number of fixed points and so on.[5]

Failures in recognition consist in the inability to find such invariants. For example, perceptual recognition can be impaired in patients with apperceptive visual agnosia who cannot recognize known objects in different contexts. They have difficulty recognizing common objects viewed from unconventional angles but have no difficulty identifying objects shown in conventional orientations (Warrington and James, 1988).

The crucial point is to explain what these invariants are. The notion of invariance was introduced into the theory of depiction by Gibson (1971) to explain how pictures can convey the same optical information as in perception. The concept of optical information covers the idea that when one sees an object, one does not see only its front surface but the whole of it: the back and the front. In a sense, all of its aspects are present in the experience.

According to Gibson (1973), perception is a matter of identifying formless and timeless invariants that specify the object's distinctive features. We can think about perceptual invariants in the same way as we think about geometrical objects. The form of a geometrical figure, like a triangle, is the face of the figure. While the face can change during transformations, the substance of a triangle remains the same.

A picture is an array of persisting invariants of a structure. They can be perceived not only when the perspective keeps changing, as in ordinary perception, but also when it is arrested, as in pictures. This means that when I see a cat, I perceive an invariant of the cat. When I encounter a picture of this cat, I am prepared to pick out the relevant invariant. This is not to say that I see an abstract cat, I have that-perception of a cat, or I perceive common features of the class of cats. What I get is information on the persistence of structure.

A picture is a record of perception. It enables the observed invariants to be stored. These invariants can be retrieved to convey knowledge. Thus, pictures are an efficient method of teaching and learning. However, the knowledge pictures convey is not explicit and language-like. According to Gibson (1978, 1979), the formless invariants cannot be put into words. They can be captured but not described.

However, the invariant concept is unclear (Fodor and Pylyshyn, 1981; Gombrich, 1971). Particularly, it is not clear what it means that some invariants can be perceived when the perspective is arrested. Let us consider the projective geometry case. In his early account of depiction, Gibson (1954) defends the projection-based theory, according to which a picture conveys optical information through the geometrical projecting of three-dimensional scenes onto two-dimensional surfaces. In short, picture P depicts object O iff there is a systematic mapping between P and O such that it preserves some properties determined by the projection invariants. Based on the principles of projective geometry, interpretation of P is a matter of identifying the scene that was projected onto the picture.

The projection-based theory seems to fit well with satellite maps, photographs and Alberti's paintings. However, Gibson (1971) explicitly rejects this theory. First, it cannot explain distorted pictures like Cezanne's paintings. Second, it is at odds with depiction practice. Contrary to popular belief, artists have never paid much attention to projective geometry clarity and abstract elegance. For instance, they commonly confuse the habit of putting the vanishing point in the centre of the picture, which is a matter of composition, with the perspective projection system as such (Gombrich, 1972). Third, the ability to recognize depicted objects in the flesh survives significant changes in the properties of the representation *and* the represented object (Lopes, 1996). For instance, we can reidentify depicted objects, even if the depiction does not represent the perspective correctly and the depicted object has changed over time. This implies that the idea of projective geometry is too restrictive to explain recognition (Inkpin, 2016).

However, projective geometry gives us an insight into what invariants are. Let us consider the cross-ratio invariant represented in Figure 5.2.

The cross-ratio remains unchanged across projective transformations. It is preserved regardless of the shape of the projection plane. It is the invariant and

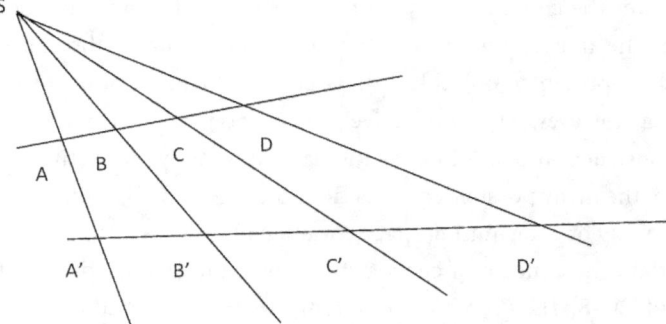

Figure 5.2 The cross-ratio invariant. The relation between points is preserved in projective transformations in such a way that the cross-ratios $AC / BC : AD / BD$ and $A'C' / B'C' : A'D' / B'D'$ are equal. © Piotr Kozak.

the property of the projection system. However, it would be wrong to say that it does not describe the properties of the depicted object. It certainly does. Based on the lengths of $A'C'$, $B'C'$, $A'D'$, $B'D'$ we can deduce the lengths of AC, BC, AD, BD.

The main idea is that images are representations of invariants. We can observe the invariants in the projection plane and apply them to the represented reality (e.g. Elgin, 2010a). In this sense, we see invariants in images. How should we understand this?

The concept of construction is useful here. The content of diagrams has been provisionally defined in terms of representing construction rules. A successful interpretation of a diagram is based on identifying construction parameters determined by these rules. The same definition is extendable to all image genera.

Constructions have been characterized in terms of determining the parameters of some space in order to arrive at the constructed object. By determining these parameters, we determine the properties of some informational space. The general idea is that some properties remain unchanged across different constructions. These properties are the construction invariants. They determine what is being preserved during different acts of determining parameters of a space.

Construction invariant: the property that is identified as unchanged in the act of construction.

Let me go back to the projective geometry example. If we apply the rules of projective geometry to construct the line segments $ABCD$ and $A'B'C'D'$, the cross-ratios are preserved. The cross-ratio is the construction invariant. It means that whatever the construction parameters, such as the vanishing point or the lengths of the line segments, are, the cross-ratio remains the same. We can identify the represented object length based on the lengths of the projected one.

The idea, however, is that projective geometry is just a special case of the more general construction operation. As Riemann might say, projective geometry is only one of the many possible constructions. In other words, projective geometry is only one way of determining space parameters.

To explain the concept of construction invariants, let us consider the terms '7 + 5' and '20−8'. They represent different constructions that identify different parameters of the logical space. However, these constructions pick out the same target. The number 12 is no less than the invariant of different constructions (while not being a projective invariant). We identify 12 by finding that this value remains unchanged in various constructions.

Next, think about triangles different in shape and size, represented by two diagrams. The diagrams exemplify different constructions. However, these constructions preserve some properties, such as the number of angles. Interpretation of these diagrams is based on identifying these construction invariants. Based on that, we can identify the constructed objects as belonging to the same kind.

Depending on the identified construction invariants, we can recognize different targets. For instance, if we construct a triangle whose invariant is the sum of its angles, then we identify different shapes of triangles as representing one kind. If the construction invariant is the angle measure, then we identify two triangles with the same angles as similar.

Cezanne's paintings present a more demanding case. To explain it, let me go back to the Gibsonian concept of perceptual invariance. It describes a familiar phenomenon of perceiving objects as stable throughout environmental changes. For instance, perceived colours seem relatively constant under changing illumination conditions. A red apple looks red at midday and at sunset. This subjective constancy of colours helps us to identify objects and is indispensable in object recognition.

According to Gibson (1979), the (approximate) invariance of the colour experience cannot be a matter of identifying the perceiver-independent properties of a perceptual object, such as the light wavelength. These properties differ throughout the changes in illumination and perspective. Instead, colour constancy is a matter of tracking relational features of the perceptual object. It is a skill of identifying the perceived colour in relation to other colours across contexts (Buccella, 2021; Green, 2019).

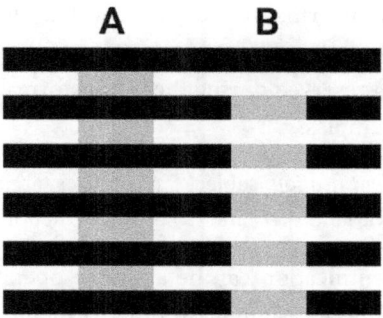

Figure 5.3 Munker-White's Illusion. Although the grey bars A and B depict the same hue, B appears to be brighter. Source: Wikimedia Commons.

A good illustration of the relational nature of colour constancy is the so-called Munker-White's illusion (Figure 5.3). The grey bars, A and B, depict the same colour. However, we perceive them differently, depending on the colour of the bordering stripes. Our perception of colours is partly determined by the surrounding colours.

According to the relational colour constancy hypothesis (Craven and Foster, 1992; Davies, 2016; Foster, 2003, 2011), recognizing the perceived colour by the perceptual system cannot be based simply on picking the colour out. The system has to contextually identify the relations between different colours to find the searched one. It must track the colour of the target surface by identifying the relations of this colour to the adjacent surfaces as the illumination changes. In other words, the system has to identify the parameters of some space of colours, such as the colour of the adjacent regions surface, and learn how they change across different perceptual contexts. The searched colour is invariant across different parameters of the perceptual space of colours, including changes in illuminations and perspective. The colour constancy mechanism is based on keeping track of invariant relations among adjacent surfaces as the perceptual conditions change.

What does it teach us about Cezanne's paintings? One of the impressionists' main goals is to depict how we perceive colours. To do so, Cezanne, using paint, determines the parameters of the depicted colour space to find the searched colour. He picks out some colours and identifies relations between them, determining the construction parameters. He achieves the goal if such constructed colour space preserves the same colour as in perception. The colour experience is the invariant of this construction.[6]

I hold that recognition-based identification consists in identifying the construction invariants connected with knowing the construction rules that determine possible transformations of the constructed object. Recognition-based identification means knowing what the permissible transformations of the object are and what is preserved during these transformations. Let me explain it starting from some examples.

> *Recognition-based identification*: subject S recognizes target T, iff (i) S identifies the construction's invariants of T; and (ii) the construction rules of T.

Recognition-based identification can be a simple operation. Suppose you are looking for a colour hue for your bedroom. To identify the searched colour you can take a colour chip (note that it is not a projective geometry case). We can change the chip's size and shape. However, the colour property has to be preserved. Based on the chip, you can recognize the searched hue. The chip is the representation of the colour invariant by means of which you identify the colour.

Similarly, to identify the way we experience colours, you can look at Cezanne's paintings. Although they do not preserve shapes and distances, they tend to capture how the colours are perceived. The colour experience is the construction invariant. Cezanne's painting is a representation by means of which you identify how one perceives the colour.

Notably, recognition-based identification is directly referential. It is not a matter of knowing an object's description and matching properties. To know the reference of the construction 7 + 5, you do not have to compare 12 to some mysterious searched value, looking for similarities between them. You directly identify 12 as the correct answer because you know how to count.

Perceptual recognition is a skill of identifying the parameters of some perceptual space – basic representational units, like edges and colours, and their relations – and transformational principles. Recognizing that perceptual objects belong to the same kind is a skill of identifying invariants that order some perceptual manifold. Invariants are the means to identify structures and objects in perception (Ison and Quiroga, 2008; Lowe, 2004).

Images are representations of these invariants. Images do not show objects. They show how to localize them by identifying construction invariants. They show how to see objects.

Comparing it one more time with the structural theories of depiction can make it more clear. According to structural theories (e.g. French, 2003; Kulvicki, 2014), images are structure-preserving representations that are structurally similar to the represented objects. Let us recall the following definition of structural similarity: two sets, *A* and *B*, are structurally similar iff there is a function mapping (some) elements of *A* and (some) relations defined over the members of *A* onto (some) elements of *B* and (some) relations defined over the members of *B* such that preserves the second-order properties of the relations in *A*. For instance, sets *A* and *B* are structurally similar if they are isomorphic without necessarily sharing any first-order property.

Now, structural theories of depiction are not wrong. Moreover, the theory of construction invariants can be seen as a version of a structural theory. A structural correspondence between the representation and represented object is another way of saying that some object properties are preserved by the representation. For instance, a map identifies a territory only if there is a structural correspondence between the map and territory. However, the structural theory heavily depends on how we understand the condition 'some'. If we want to find these elements and relations, we need a defined domain, range and mapping function. It is no problem in the case of mathematical operations. It is a problem, however, if we want to map a defined mathematical structure onto an undefined set of states of affairs. Structures are not some Platonic objects waiting to be discovered in reality and compared. It is not sufficient to say that semantics is fixed or that some classes or mappings are more natural. It is nothing more than a metaphysical postulate (Isaac, 2019; van Fraassen, 2008).

The image goal is to mark the elements and relations to impose order on some manifold and identify the structure. Images determine the properties of the informational space to localize the searched object. Successful interpretation of an image is based on finding the construction parameters (together with knowing construction rules) that help us to localize this object. For instance, the triangle diagram highlights the construction parameters, like the number of angles, in order to identify some objects as belonging to the same kind. Understanding the diagram implies that you know that you can change the diagram's size and shape (but you cannot disconnect the elements of the structure), and some properties are preserved. It does not involve comparing it to some ideal Platonic object. It involves knowing how to identify triangles by learning what remains unchanged across different constructions.

Let me illustrate it with the following example (Figure 5.4). Suppose you have some manifold represented by a set of points (Figure 5.4a). Images impose order on this manifold by determining the construction parameters: a relevant set of points and order relations (Figure 5.4b). Interpretation of the image is based on identifying these construction parameters. Knowing that this construction preserves the number of angles, we can identify it as representing a triangle (Figure 5.4c).

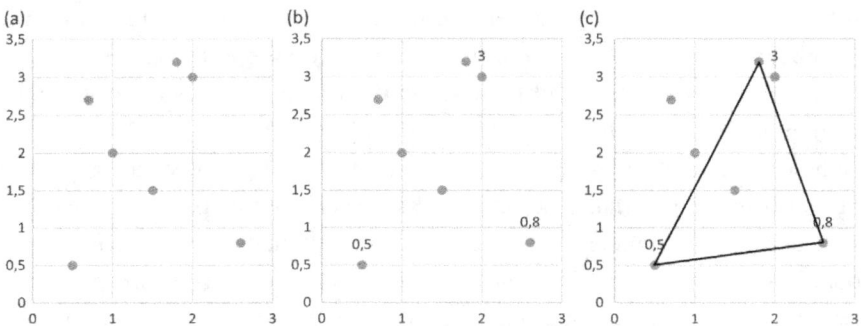

Figure 5.4 Images impose order on some manifold to localize the elements of the structure. © Piotr Kozak.

Are construction invariants the properties of the representational system or the represented object? It is a misplaced question. Recognition is the skill of identifying the construction invariants to localize the searched object. Construction invariants are the ways of describing the objects, not the descriptions of an object. They show how to localize an object. Images exemplify the construction invariants by means of which you identify some properties. What does it mean?

Let us think about construction invariants as measurement invariants. Measurement invariants are these properties of the measurement that indicates that the same construct is being measured. For instance, the standard metre length exemplifies the measurement invariant by means of which we measure length. No matter what the world is like, the standard metre will be one metre long. Does the standard metre describe the world? In the most general sense, it is. There has to be a structural correspondence between the length of the standard metre and the measured length. However, it should be understood as saying that the length of the standard metre is taken to localize the properties of the world. The standard metre represents the way of localizing the world's properties.

Now, you cannot ask how you know that the standard metre is one metre long, for it sets the condition of identifying what one metre is. To understand what the standard metre indicates, you have to learn what properties it preserves. It is learning a measuring convention. Nothing more. By the same token, asking how you know what construction invariants are preserved is the wrong kind of question, for they set the conditions of identification. To understand what construction invariants types are preserved is to ask about the convention they are a part of. Nothing more. Think about a colourful triangle diagram. To ask how you know that it preserves shape properties and not the colours is like asking how you know that a ruler represents a distance and not a ruler's colour.

Thus, knowing how to localize objects is not based on feature comparison or looking for the similarities between some objects. It involves knowing what properties remain unchanged during different constructions. Consequently, we can track and reidentify objects with different physical properties, for the object description can change across different ways of localizing searched objects. The invariant can be the mere property of being individuated by some construction. Let me illustrate this with three cases.

First, consider a simple case of setting a localization of some point in space. It can be done by determining the coordinates in relation to some frame of reference represented by a Cartesian diagram. However, the frame of reference and, respectively, the coordinates can be altered. Yet, when changing the frame of reference, we do not change the localization of the point. We change the description of the localization. This does not imply that the point is localized in some absolute space. Here, localization is the invariant of the way we localize the point. The Cartesian diagram determines the properties of this space.

Second, in the tunnel effect experiments, subjects are not asked to identify the properties of perceived objects. They judge that the object that disappeared when moving beyond the occluding shape and the one that reappeared on the other side are one and the same even if all object properties have changed. Thus, identifying likenesses is not necessary for recognition-based identification. In the case of the tunnel effect, identifying the relation to the vanishing point and the motion vector as a transformation invariant is sufficient to recognize these objects as belonging to the same kind.

Third, the represented properties do not have to be 'localizable' in the image (Wollheim, 2003). I can see that a depicted person is happy even though no single property of the image can be identified as representing happiness. Happiness is the depicted object's property localizable by identifying construction invariants. I can see-in the depicted person's happiness by identifying patterns of representation, for example, the composition of colours and lines in a painting. These patterns are the construction invariants represented in a picture through which we identify the represented object.

Let us go back to the black hole picture. Its interpretation involves determining the construction parameters of the black hole. It is based on isolating the object from the background by determining the parameters of some space. Recognition-based identification of the black hole means identifying the pattern of the black hole. It is not based on comparing the picture's properties to the black hole, which is physically impossible. It is about knowing which properties of the construction of the black hole remain unchanged. For instance, determining the black hole's spatial properties means identifying the emission ring's crescent shape and the central shadow. These are the invariants of the black hole construction and the means to localize the properties of the black hole. These properties are close to the construction properties of a donut. That is why it is easy to confuse the black hole picture with the picture of a donut. Yet, no one says that knowing the interpretation conditions implies that we will always arrive at the correct one.

Iconic convention

Recognizing construction invariants is not an easy task. It requires expertise, as in the case of knot diagrams and fMR images. The ability to recognize such invariants is a matter of skill. Yet, when it is mastered, it enables us to interpret different constructions that preserve the same invariants.

Consider the triangle diagram. If you learn how to identify invariants in one triangle diagram, you learn how to identify them in any other. You learn a convention of how to identify certain invariants. Learning this convention implies that you know how to go on, in the same way as knowing how to count implies that you know how to count to infinity.

Accordingly, iconic convention can be defined as a set of construction rules that preserve the same type of construction invariants. Two pictures belong to the same convention if the same invariants types are identified.

> *Iconic Convention*: a set of construction rules distinguished by the same construction invariants types.

For instance, projective geometry is a set of construction rules that preserves cross-ratios. Distances are preserved in isometric transformations. Photographs preserve optical information and light direction. Impressionist paintings preserve the visual experience of the object and so on.

Knowing how to interpret images is knowing the iconic convention determining the construction invariants type. For instance, reading a topographic map is based on the ability to identify the information on distances and spatial relations between locations. Distances and spatial relations are construction invariants of the topographic convention. Alternatively, reading a topological map is based on identifying topological relations, not distances. Topological relations are invariants of topological constructions.

The fact that there are many iconic conventions is not accidental. If constructions are characterized in terms of determining the parameters of some informational space, then different constructions can highlight different states of the space. Iconic conventions can be compared to different systems of reference that identify different invariants. If there is no absolute point of view, then there can be no one iconic convention.

Interpreting iconic convention in terms of construction invariants types helps us to address the problem of natural generativity. The problem is how to explain the ability to interpret a whole system of representation based on the meaning of its part. Applying the concept of the iconic convention makes the concept of natural generativity explainable and avoids the problems of recognition-based accounts of depiction. Consider a case of a mathematician who learns how to interpret a graph of the function $f(x) = x + 1$. He learns the graph construction rule. Knowing this, he can recognize other functions, such as $f(x) = x + 2$, displayed on a graph, too. This seems to be the case of natural generativity. However, there is no need to be perceptually acquainted with any function. It is sufficient to know what properties it preserves.

Consequently, we can address the problem of the naturalness of icons. According to the metaphysical understanding, the representational system is natural only if the properties of the system depend on the properties of the represented object. In this sense, iconic representations are natural. For instance, if the distances between represented objects were different, then the distances displayed on the projection plane would have to be different, too, for the cross-ratio is preserved in projections. If Cezanne had a different visual experience, then the visual properties of his paintings would have to change, too, for the experience properties are preserved in impressionist paintings. In contrast, symbolic representations are not natural. Whether the symbol A represents the left or right direction is an arbitrary choice.

In contrast to recognition-based theories, knowing an iconic convention does not involve perceptual acquaintance with the represented object. It is sufficient to know which properties are preserved by the iconic convention. This seems justified since we commonly depict objects we have never seen before. For instance, we have no problems depicting fictional objects like dragons. We do not have to see dragons in the flesh. Recognition-based identification does not involve comparing two relata: the depiction and the depicted object. It is sufficient to know how a dragon would look if it existed. If we know what properties are preserved by the iconic convention, we can infer what kind of properties dragons would have.

Let us go back to the black hole picture. Recognizing the black hole in the picture does not require that we stand in some perceptual or resemblance-like relation to the black hole. If we know the iconic convention of the picture, we know which properties of the black hole it preserves. That is why we can infer the properties of the black hole from the picture.

Consequently, the skill of recognition can be dissociated from having descriptive knowledge. Let us suppose that the picture of a black hole resembles a square rather than a circle. Still, knowing the iconic convention that optical information is preserved in a photograph, we could recognize it as a black hole picture. The square-like black hole would not correspond to our common way of representing black holes, but no one says that science should align with how we commonly represent things.

Summary

This chapter discusses the recognition concept and its relation to images. Recognition is a natural and direct relation to the represented object. I hold that it is a distinct kind of reference relation that is irreducible to demonstratives and descriptions. Images are necessary to represent this reference relation.

I explain the recognition in terms of identifying the construction invariants, that is, the properties that remain preserved in different constructions. I hold that recognition is based on identifying these invariants. In the last section, I explicate the concept of iconic convention, characterized as a set of construction rules that preserve the same invariants.

6

What is an image?

In the previous chapters, I characterized images' functions as necessary for representing operations of construction and recognition. In this chapter, I present a two-dimensional model of reference for images. I hold that the best way to understand the properties of this model is to think of images in terms of measurement devices, such as rulers and balances. I hold that images show how a measured object would look in some measurement set-up by representing the ways of identifying this object. Finally, I apply the measurement-theoretic account to explore the nature of impossible images.

Two-dimensional model of iconic reference

Before characterizing the model of iconic reference, some conceptual distinctions are needed. Drawing on Cummins (1996) and Greenberg (2018, 2021),[1] I introduce a distinction between the referent, content and the representation target.[2] The target is the object or the state of the world the representation is aimed at. The referent is the object or the state of the world that is actually identified by the content. The content are the properties of representation through which we identify the referent. The properties of the content determine the referent. The target can be fixed even before the representation is created.

Distinguishing between the target and referent is essential for making sense of the concept of accuracy conditions. The general idea is that a referent can mismatch a target. For instance, I intend to draw a horse, but because of my lack of skills, the drawing resembles a cow. A cow is a referent identified by the content, and a horse is the target of this representation. If the referent mismatches the target, an image is inaccurate. It is accurate otherwise.

Images can represent the same target but have different content. Suppose my seven-year-old son's drawing and Alberti's paintings depict the same targets. Yet, they depict them in different ways, having different content. There can be cases when pictures have the same content but different targets, too. The same picture of John can represent John, as well as a man as such if it figures in a biology textbook.

Seeing Marilyn Monroe on black canvas (Schier, 1986) is a special case when we identify a target without matching content. We take Marilyn Monroe to be a target, but the content of the black canvas is not Marylin Monroe. The Putnam's ants case (1981) is the reverse of that. A Churchill-like trace left by an ant does not refer to Winston Churchill. He is not a target of this representation. However, the content of the trace identifies the properties of Winston Churchill. There can be targetless representations that have content, too. Impossible images aim at no target. Still, the impossible fork and the Penrose stairs have different content.

In depiction theories, these dimensions are often conflated. Wollheim's (1987, 1998) seeing-in theory fails to distinguish between the content and target. Consequently, it cannot identify correctness standards that determine whether the viewer has correctly perceived what a picture represents. Wollheim suggests that the author's intentions can do the job (Wollheim, 1980, 1993). However, the concept of intention can be applied to identify the target but not to the content. My seven-year-old son and Alberti can have the same intentions, but they produce different content (Hyman, 2012, 2015).

The resemblance-based theories are the theories of content, not target. Resemblance can determine correctness conditions for content but is too liberal to specify a target. The reason is that, in some respect, everything resembles everything else. Goodman's famous attack on resemblance (1972, 1976) is an attack on resemblance as a theory of target (e.g. Greenberg, 2013; Walton, 1990). However, Goodman's theory is a non-starter, too, for it rejects the distinction between content and target, reducing the theory of content to the theory of target.

The inability to distinguish between content and target is a source of confusion in the theory of depiction. It suggests that the main problem is not that there is no unified theory of depiction; maybe no such theory is possible. The problem is that we have many theories that pick out different topics.

We are now in a position to characterize the iconic reference. The relationship between an image, content and target is two-dimensional. The general idea is that an image (in respect of the circumstances of evaluation) denotes a target and the properties of the carrier exemplify the properties of content. In turn, the properties of content (in respect of the iconic convention) are used to reidentify (in terms of recognition-based identification) the referent. The content indicates the target by means of reidentifying the referent. Depending on whether the referent matches the target or not, the image is either accurate or inaccurate. Let me call it the two-dimensional model of iconic reference (2-D for short). It is represented in Figure 6.1.

Let us start with the concept of an image. In the 2-D model, it is the carrier of the representational function. The carrier of a representation can be a physical object, for example, a picture or a diagram, and a psychical one, for example, a mental image. For instance, the carrier of the portrait of Duke Wellington is the canvas covered with pigments, and the carrier of the mental image of Duke Wellington is the mental representation that underlies the experience of the Duke. The main question of the theory of iconic reference is how the properties of the carrier, such as pigments on the picture's surface, can represent something outside of the carrier. In other words, what is it that turns an image into a representation of something else?

Drawing on Goodman (1976), I distinguish between denotation and exemplification as two kinds of reference relations. According to the 2-D model, images denote the target and exemplify the content. Denotation is a dyadic relation between a representation and something it stands for.[3] Exemplification runs in the opposite direction. It refers back from the selected properties of the object to its representation. Additionally, it requires instantiation. A representation has to have the exemplified property. For instance, a water sample exemplifies the water quality, for it instantiates it. Exemplification is instantiation plus reference.

However, not every instantiated property is exemplified. A water sample instantiates the property of being taken on a certain date. Yet, the sample does

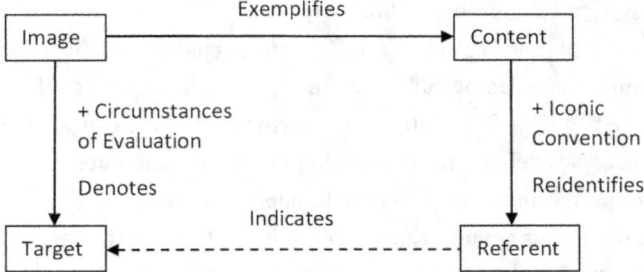

Figure 6.1 The 2-D model of iconic reference representing the relationship between an image, its content, the referent and the target. An image denotes a target in respect of the circumstances of evaluation and exemplifies the content, which reidentifies the referent in respect of iconic convention. The content indicates the target by identifying the referent. © Piotr Kozak.

not exemplify the date. Depending on the context, only selected properties are exemplified.

The role of a sample is to highlight the selected properties of the object. To do so, the highlighted properties of the sample have to be salient. A bald man wearing a hat is not a good sample of baldness. A sample has to be filtered (Elgin, 1983, 1996b; Goodman, 1984; Goodman and Elgin, 1988).

Let me explain the nature of iconic reference, starting with iconic denotation. There are two general approaches to define it (Abell and Bantinaki, 2010). On causal models, images denote their objects in virtue of being part of a causal chain between the representation and the represented object. In intention-based models, images denote in virtue of the author's intentions which aim at specific objects.

The intention-based approach works well for paintings and drawings. Yet, it hardly applies to photographs, for they are based on causal rather than intentional relations. Causal models do not apply to fictional representations. There is no causal relation between a unicorn and its picture, but there are pictures of unicorns.

To find a way out, we need to make two remarks. First, these two approaches are often conflated. Let us recall Kripke's causal theory of proper names (1980). In order to identify the denotation of a proper name, you have to fix the denotation in the act of the initial baptism. However, the initial baptism is not part of the causal chain, which is fixed intentionally. The causal chain is the way we identify the intention-based act of baptism.

The general idea of the intention-based approach is that intention carries a teleological function. A intends B in the sense that B is a goal of A. Causal relations can be easily wired into such teleological functions. For instance, some causal chains can be picked out by carrying some teleological function in a biological system (Cummins, 1996).

In the case of photographs, certain causal chains can be picked out by distinguishing the teleological function of the photographs. If a picture is distorted due to a lens defect, the lens defect is part of the causal chain, but it is not the target. We exclude this causal chain from the reference relation because the photograph has not been designed to depict lens defects.

Second, intentions are not transparent. It is not the case that, when producing an image, we have a clear view of our intentions, as if some transparent mental events guide our actions. Rather, we infer our intentions by judging the final product of our intentional act. Architectural sketches well illustrate this. If I start sketching a building, I do not have to know what I want to draw. I am searching for it.

Moreover, depending on the knowledge of accompanying causal factors, we can change the interpretation of the target. For instance, if we learn some facts about an author, we can reinterpret his paintings (Hopkins, 1998; Lopes, 1996; Terrone, 2021a).

To reconcile the intention-based and causal theories, let us think of iconic denotation in terms of the measurement theory. In measurement, the target is the measured quantity. The measured quantity is the state of the logical or physical space determined by different locations of information within this space. For instance, 12 is localized within the logical space by its position in the ordered numerical set. The objects in the physical space are localized within the spatiotemporal frame of reference. The goal of the measurement is to identify these states. Measurement refers to a measured quantity that it was designed to identify.

By the same token, images denote their targets in terms of identifying some states in the logical or physical space. An image denotes its target if it is aimed at identifying these states. For instance, the inflation diagram denotes inflation growth, and the black hole picture denotes the black hole, for they were designed to depict them. In contrast, the Churchill-like trace left by an ant does not denote Winston Churchill, for it was not designed to depict Winston Churchill.

The image target is independent of the content. Depending on the goal, the same photograph can be used to depict a black hole or to test measurement devices. In the same way, the same ruler can be used to represent straight lines and distances. Interpretation of images is partly a matter of identifying these goals.

Such interpretation is an ongoing process, for there are cases when an image shifts its reference. We intend to portray Jane, but instead, we portray Sarah, her twin sister. Before we knew that Jane had a twin sister, the picture denoted Jane. After we learned that, the picture's target is Sarah. By the same token, a thermometer can be used to represent temperature. Yet, by learning thermophysics, we can interpret it as representing pressure. Identifying the target of a depiction is theory-laden and depends on our knowledge of the world.

At first glance, the measurement-based account of iconic denotation can be seen as a refined version of the intention-based theory since identifying the picture's target can be based on identifying the intention of the picture's producer. However, there are significant differences. The measurement-based account does not posit the existence of any mental event for identifying the relevant state. A state can be identified by the use of an image in some epistemic practice. An image is of something if it is used to identify the relevant state of space.

The source of confusion is that the concept of intention is nested within our cognitive practice of explaining representational content. Ascribing intentions can be seen as taking an intentional stance (Dennett, 1987) to explain the object of representation. It can be useful to describe an image in terms of mental events, such as intentions, to identify a targeted state. Yet, intentions are only a method of identifying these states. The concept of intention is like the concept of a meridian. It is used to identify some world properties but is not the same object as a chair.

Does this mean that intentions are not real? This is a misplaced question. It is like asking whether a meridian is real. These are certain ways of describing the world, but they are not a description of the world. Depending on the level of explanation, the intentional stance can be replaced by the design stance or the physical stance. However, taking the design or physical stance does not falsify the intentional stance. It is only a matter of the way we describe the world.

The measurement-based account can be reconciled with causal theories in the following way. Identifying the target enables us to pick out relevant causal factors. The black hole picture denotes the black hole, for it is part of a causal chain whose initial element is the black hole. At the same time, the picture has been designed to depict the black hole. Depicting it was a goal of the picture. Therefore, irrelevant causal links, such as distortions caused by the defects of measurement devices, are excluded from the denotation.

Yet, the existence of the causal relation is neither necessary nor sufficient for denoting the target. The unicorn picture identifies a possible world inhabited by unicorns and picks them as its target.[4] Yet, there is no causal link between them.

However, this does not imply that causality plays no role in determining iconic denotation. There are images, such as unicorn pictures, that denote objects in some possible worlds, and images, such as black hole pictures, that denote objects in the actual world. The key is to identify what possible world we are in. The condition of identifying isomorphism between the properties of a possible world and an image is too weak. A map of London can be isomorphic with Mordor, but Mordor is not part of the actual world. Looking for causal correlations seems to be a good criterion for such identification. Picking out the causal relations between an image and the target indicates that the possible world overlaps with the actual one.

Following Kaplan (1989), identifying the state of space and the possible world we are in establishes the so-called circumstances of evaluation. It can be defined as a function from the image to the possible world. By identifying the possible world, an image can locate a target in a possible world w_1 and be targetless in w_2. Compare a documentary and a feature movie. The target of the first is located in the actual world; the target of the second is in a fictional world (Terrone, 2021b).

To sum up, we arrive at the following definition of iconic denotation.

Iconic denotation: Image *I* denotes target *T*, iff (i) *I* is designed to represent *T*; whereby (ii) *I* identifies a targeted state of a physical or logical space; in relation to (iii) the circumstances of evaluation.

Let us turn to iconic content. In the previous chapters, I introduced the approximate definition of diagram content. It has been described as representing the construction rules allowing one to get to the searched object. The same definition can be extended to the content of images. An image exemplifies the iconic content to reidentify the referent by following the iconic convention.

Iconic content: An image *I* exemplifies content *C*, iff (i) *I* instantiates and represents *C*; (ii) *C* identifies the construction rules and (iii) construction invariants; whereby (iv) *C* reidentifies the referent *R* in relation to (v) the iconic convention.

The general idea is that iconic content exemplifies the construction rule, which determines the parameters of some informational space necessary to identify the referent. Iconic content shows the ways to localize some state of physical or logical space. We can unpack this definition based on the concepts of recognition-based identification and iconic convention.

Primarily, an image does not exemplify the properties of the target. This restriction is necessary, for, without it, the 2-D model would be vulnerable to objections of the following kind. For one thing, exemplification could not ground the content, for not every property can be instantiated. A picture of my mother represents her as happy, but the picture is not happy. Similarly, there is no way to exemplify the properties of unobservable phenomena. An image of the Higgs boson does not instantiate any property of the boson (e.g. Frigg and Nguyen, 2020; Mößner, 2018). For another, the symbol 'word' refers to a word and exemplifies the property of being a word. Yet, it is not a picture of a word (Goodman, 1988).

We can easily avoid these objections, for an image exemplifies the properties of constructions, not the constructed objects. The properties of construction and the constructed objects should not be confused. Consequently, images can exemplify unobservable phenomena, as they do not show them. They show how to localize them in some physical or logical space.

However, not every carrier property is a property of the content. Only these properties of the carrier that are highlighted by the construction parameters are relevant. For instance, I can draw a square imprecisely because of my lack of skills. Still, I recognize it as a picture of a square, for it represents the construction rules of squares. The content highlights relevant properties of the representational carrier.

Notably, iconic content does not represent objects. It represents the ways of identifying objects. Interpretation of iconic content is based on identifying construction parameters determined by the construction rules. These parameters are the basic units of construction, their order and the relation between them. Construction rules determine the permissible transformations of these parameters. Based on the iconic convention, we can identify construction invariants, that is, the properties that remain unchanged across different constructions. Finding these invariants brings about the recognition-based identification of the referent. The referent is a property, object, event or scenario identified by the content.

The construction parameters of iconic content are representational units, such as colours and lines, and relations between them. They have to be distinguished from the registration of the representational content. For instance, brush strokes on the picture's surface register the colour, but the representational unit is a colour, not a brush stroke. Finding how to register content is a matter of representational technique. In the same way, it is a matter of measuring technique to use mercury (instead of, e.g., alcohol) to represent the temperature in a thermometer. However, the representational unit is the height of the mercury column, not mercury.

Consider a picture of John. The target of representation is John. However, the way it is represented, the composition of lines and colours, is the content of the picture. The referent is the object identified by these lines and colours. Brush strokes are the ways colours and lines are registered.

The referent of iconic content can be either a particular or general object. Let us recall the distinction between token- and type-diagrams. A token-diagram represents a particular token, such as a specific triangle. A type-diagram represents a general object, such as a triangle as a such. Depending on the identified construction invariants, every image can be taken to represent either a particular or general object. However, every type-representation involves representing a token (but not vice versa).[5]

Importantly, recognition-based identification is directly referential. It implies that iconic content is identified directly in the picture. It is not inferred from the picture. However, this does not mean that representations have a mysterious feature of being intentional in terms of seeing a nonphysical *Bildobjekt* (Husserl, 2005) or that we are seeing-in the represented objects (Wollheim, 1987). These objects are artefacts of the measurement operations. How should we understand this?

Consider the measurement operation. Saying that the term 20°C indicates the particular temperature is not saying that we are intentionally directed to some abstract number. This number is a construction property by which we localize some state of the physical space determining its relations to other states, such as 19°C and 21°C (Armstrong, 1989; Swoyer, 1987). Saying that the temperature is 20°C means that this number localizes some particular state of the physical space. In this sense, we can directly identify this state by this number.

Let us compare it with iconic denotation. It has been characterized in terms of dyadic relation to the target. In contrast, iconic content is not a relation to some abstract referent. It is a way of localizing the referent. Localizing a referent means finding its location in some physical or logical space, in the same way as the number 20°C directly localizes some particular temperature value.

Images appear intentional, for we use intentional vocabulary to describe them. For example, I can say that I see my mother in the picture. However, saying that I can see my mother in a picture is not saying that I can see an abstract idea of my mother. In the same way, saying that the temperature is 20°C is not saying that the temperature has some numerical content. Iconic content, just as 20°C, is the way to localize a certain state of the world.

Exemplifying content is insufficient to recognize the referent, for the representational carrier can instantiate many different properties. Content requires interpretation following the iconic convention. The goal of interpretation is to select the properties that contribute to the recognition of the target. Interpretation highlights these properties by identifying construction parameters.

Interpretation of content is ruled by the Davidsonian-like principle of charity. The principle holds that selected construction properties are exemplified to bring us closer to the target. In other words, we presuppose that the goal of content is to identify the designed target.

However, the nature of the principle of charity is heuristic. We can take different interpretational stances and shift the target. For instance, we can abstract away from the author's intentions and study the social factors that are causally manifested in a picture. In this sense, the picture's interpretation is free of the author.

Thus, we arrive at the following definition of the 2-D model of iconic reference.

2-D model of iconic reference: image I refers to target T, iff it (i) I iconically denotes T; and (ii) I exemplifies iconic content C; and (iii) C identifies referent R; and (iv) C indicates T through identifying R.

Importantly, matching the referent and target cannot be based on comparing features of the target and referent; for then, we would have to reintroduce descriptive content. Let us think about this matching in terms of the indexical relation between the content and the target. The content indicates the target by identifying referent in the same way as indexicals 'here' and 'now' indicate the properties of space and time. How should we understand such a relation?

Drawing on van Fraassen (2008), let us compare this to identifying the reference of the measured value. Saying that the thermometer shows 20°C is not enough to say that there is a measurement relation. It could be simply coincidental that the thermometer shows the number 20. There must be some direct link between the measure and measurand. However, saying that the measured value of 20°C matches the temperature is not holding that there are some similarities between the number 20 and the temperature. To understand this proposition, we have to know how to localize this value within the physical or logical space. It is non-propositional knowledge of how to identify the property of being 20°C within the measured space.

To locate an object in the logical or physical space is to determine which of the possibilities it realizes. It is finding a particular localization in an array of possibilities this space marks out. To locate the value of 20°C within the physical space is to find the particular localization in an array of possible states of the world. The result of 20°C indicates this localization.

Let us think about content in terms of representing a measure by means of which we indicate a state of some space by identifying the parameters of some search procedures. The content highlights certain properties to localize the target. For instance, a portrait of John highlights certain properties by means of which we identify John. These properties can be simple, as a composition of lines and colours, and complex, such as being bald. The properties of the content are like measurement parameters by means of which we identify the measured object.

Let us consider a black hole picture. Based on our knowledge of the world, we can localize a target of the black hole picture. It can be represented as a particular state of the physical space represented by a set of possible states of affairs. The goal of iconic denotation is to locate this set. Knowing the content and iconic convention of the black hole picture, we can identify the referent. Recognizing that the target matches the referent is knowing that they both localize the same state in the physical space.

Maps illustrate this point clearly. The target of a map is a state of some physical space, for example, London streets. The content of a map, for example, the composition of directions and points, identifies the ways of localizing these streets on the map. Knowing that the map preserves spatial relations, we localize a street layout. This is a referent. However, to use the map to get about London, we must locate ourselves on the map, that is, we must recognize that the map's street layout localization indicates the same localization in the physical space, that is, London. We use the referent to identify the properties of the target. A map is correct if the referent and target mark the same localizations.

Importantly, in contrast to the identification of a target, identification of the referent is not relativized to possible worlds. Consider a cross-ratio invariant. A geometric projection preserves cross-ratios in every possible world. No matter what kind of space is projected onto the plane, whether we are in the Euclidean 3-D space or in the flat world, the cross-ratio is preserved in the projection. By the same token, the triangle invariants, such as having three sides, are identical in all possible worlds. The idea is that a referent localizes the same objects across all possible worlds.

Consequently, depending on the kind of reference we pick out, an image refers to the represented object rigidly or non-rigidly. From the perspective of content, an image behaves like a rigid designator. From the perspective of denotation, an image is a non-rigid representation. For instance, in a possible world w_1 a map of London denotes London, while in a possible world w_2 the same map denotes Mordor. However, in both worlds, the content of the map is the same. Iconic content rigidly identifies the referent in the same way as the standard metre identifies the same distance in all possible worlds.

Correctness conditions

The distinction between the iconic content and target expresses two kinds of image correctness standards. In general, these standards are accuracy conditions. Accuracy is a matter of degree. An image can be more or less accurate. However, accuracy is a combination of two different kinds of standards. For one thing, an image can be a true or untrue representation of the target. For another, iconic content can be precise or imprecise. An image can be true but imprecise, and

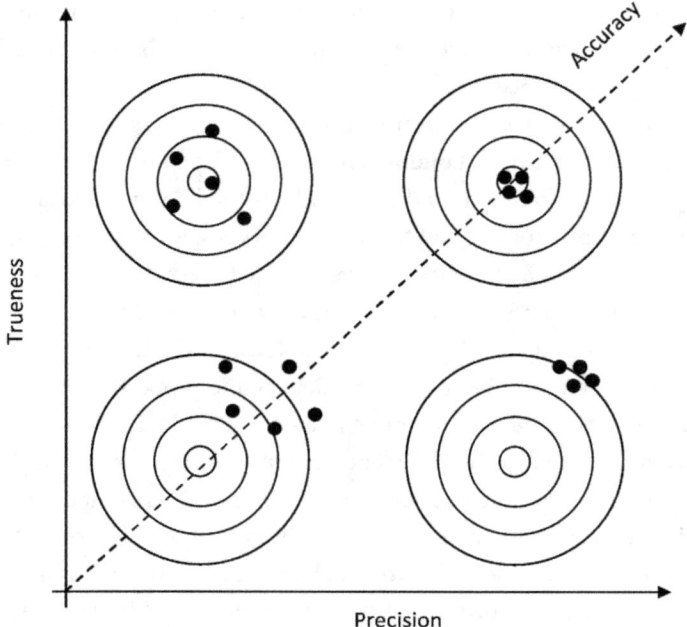

Figure 6.2 The relation between trueness, precision and accuracy conditions. A representation is accurate iff it is both precise and true. © Piotr Kozak.

precise but untrue. It is accurate iff it is both true and precise. The relation between the concepts of trueness, precision and accuracy is represented in Figure 6.2.

The concepts of trueness, precision and accuracy come from the measurement theory. Trueness represents how close a measurement is to a true value. Precision is the closeness of the measurements to each other (ISO, 1994).

We can apply these concepts to the analysis of iconic reference. First, the general idea of the precision condition is that measurement results should be unambiguous to identify the measured quantity. By the same token, the properties of iconic content should be unambiguous to identify the referent. It implies that the construction invariants should identify an unambiguous set of properties that enables us to recognize the represented object. For instance, a photograph is sharp if it identifies the properties that make us able to recognize the represented object. Otherwise, it is unsharp. The inflation diagram is precise if it displays the data that allows localizing inflation growth. Otherwise, it is imprecise.

The precision conditions are relativized to the goal of the representation and the cognitive skills of the interpreter, which both build the context of understanding an image. An image can be too little or too precise, depending on the goal. For instance, a map has to be precise enough to identify localizations. It cannot be too precise, for then it would overlap with the territory. Depending on the cognitive skills of the interpreter, images can be understandable or not. A mathematical graph is understandable only if its interpreter has the mathematical competence to understand it.[6]

Second, a measurement is true if it matches the value of the quantity it was designed to measure. By the same token, an image is true if the referent matches the target. For instance, a picture of a square is true if it was designed to represent squares. It is false if otherwise. I am not imprecise if I intend to draw a square but draw a circle instead. I am wrong.

Importantly, we should not confuse the trueness of images with the truth of propositions. The concept of trueness should be taken in metaphysically neutral terms. We do not have to interpret it as presupposing the existence of some true values (Giere, 2006; Swoyer, 1987). Instead, we can think of it in terms of robustness conditions, that is, the concepts that introduce coherence and consistency to measurement outcomes (Chang, 2004; Tal, 2013, 2017; Teller, 2013, 2018). For instance, the standard metre is not true in terms of representing the true nature of what one metre is. It determines what one metre is. By the same token, a painting can set standards of what beauty is. It does not imply, however, that some abstract object exists called ideal beauty.

Yet, although the measurement-theoretic concept of trueness is distinct from the concept of the truth of propositions, they are linked. Let me illustrate this with the concept of a sample. A sample is true if it exemplifies the measured value. For instance, the water sample is true if its quality is close to the water quality. However, it is not guaranteed that the sample will represent it. The sample can be badly taken, or the distribution of water quality values can be abnormal.

In contrast, the truth of propositions requires that we can distinguish the atomic formula of a semantical system and provide its translation into a metalanguage that satisfies a T-schema. For instance, the proposition SNOW IS WHITE is true for it satisfies the T-schema such that 'snow is white' is true iff snow is white. By using the T-schema, we can determine the truth-conditions of compound formulas. For instance, a compound proposition of the form A AND B is true iff both A and B are true. This implies that propositions have a logical form that provides correctness conditions for such translation.

Samples cannot satisfy these conditions. They can be badly taken, but they are not true or false in terms of the truth of propositions. No atomic formulas can be distinguished, for any sample represents many properties at once. Samples have no logical form. A sample of water quality does not logically imply water quality. It instantiates water quality.

However, a sample can be a basis of a proposition in the following sense. The water sample localizes the water quality property. It provides us with information about the quality of water and can be a source of the proposition THAT WATER IS CONTAMINATED. The sample works as a justification for this proposition, too. If someone asks why I hold that the water is contaminated, I can point to the sample. Yet, this does not imply that I cannot be wrong. The sample can be badly taken, and the proposition based on the sample can be wrong.

By the same token, an image is not true or false in terms of being a true or false description of a target. Images can be properly or badly taken. At the same time, images can be true in terms of the measurement theory if the referent identified by the properties of the content is close to a target. An inflation diagram is true if it is close to the true value of inflation growth. A conventional picture of a dragon is true if it is close to the stereotypical depictions of dragons. An image of a square is true if its properties are the same as those identified by the square definition. An untrue picture is such that the referent localized by the content is far from the target.

To conclude, we get different correctness conditions depending on the kind of iconic reference.

> *Precision condition*: image *I* is precise iff the *I*'s content has such properties that allow one to identify the referent unambiguously.
>
> *Trueness condition*: image *I* is true iff the *I*'s referent is close to its target.
>
> *Accuracy conditions*: image *I* is accurate iff *I* is both precise and true.

Two remarks are in order. First, the iconic reference and correctness conditions are defined in non-mentalistic terms. They do not refer to the mental states of a perceiver or a phenomenology of pictorial experience. However, it does not imply that mental states are irrelevant to understanding the meaning of images. No one holds that. It means that a theory of how we grasp meaning does not overlap with the theory of meaning.

Second, the 2-D model heavily depends on our understanding of reference, exemplification and indexicality concepts. It cannot be otherwise, for this model is intended to be part of a general theory of meaning. The implication is that the 2-D model can be incorporated into the general framework of the theory of meaning. That being said, not all model concepts can be sufficiently explicated since this would assume that we have to clarify other concepts of the theory of meaning. No book is long enough to do that.

The measurement-theoretic account of images

So far, the measurement-theoretic framework has been applied to highlight the measurement-like properties of images. This may suggest that the comparison between measurement and images is allegorical only. However, the main idea is that images are not *like* measurement devices. They *are* measurement devices. Strictly speaking, images are a kind of measurement instrument that shows how a measured object would look in some measurement set-up by representing the ways of identifying this object. Let me start by unpacking basic intuitions.

> *Image*: a kind of measurement device that shows how a measured object would look in some measurement set-up by representing the ways of identifying this object.

First, there is a long-standing tradition of interpreting images in terms of measuring devices. For instance, in the nineteenth century, photographs were hoped to be scientific instruments comparable to telescopes (Maynard, 2017).

Second, we have good reasons to believe there is a strict analogy between measurement and semantics. In general, measurement involves exploiting a correspondence between a phenomenon and certain abstracta so that features of the abstracta represent features of the phenomenon. Assigning semantic values is like setting up a measurement system (Ball, 2018; Dresner, 2002). Moreover, it is believed that the measurement theory offers a unified account of the interpretation of language, action and mind (e.g. Dennett, 1987, 1991b; Dresner, 2006, 2010, 2014; Marcus, 1990; Matthews, 2007, 2011; Swoyer, 1987).

Third, measurement is not reducible to the numerical domain. Generally, measurements are about imposing order on a certain manifold (e.g. Finkelstein, 2003; Narens, 1985). Finding whether a couch fits into a doorframe does not have to involve assigning any number to the couch.

What is a measurement? In the most general sense, a measurement consists in a systematic application of some standard to some manifold to localize a measured value. Primarily, it involves laying out what the measured quantity or category is. Next, it provides a way of a systematic mapping of the measured value or category onto the representation. Finally, it determines what must be done to successfully carry out the measurement (Bradburn et al., 2017; Cartwright and Runhardt, 2014; Chang and Cartwright, 2014).[7]

Every measurement is theory-laden. We need a theory or a model to understand what is being measured. Every act of measurement depends directly or indirectly on the outcomes of other measurements. For instance, measuring air temperature depends on the outcomes of the measurement of air pressure.

Measurement is selective. Only selected properties are represented. It is perspectival. It does not show the measured value from some absolute point of view but in a measurement set-up. The measurement result is relativized to the measurement system. Both the Kelvin scale and Celsius scale measure temperature. However, the Kelvin scale represents thermodynamic temperature, while the Celsius scale represents the intervals between temperature values (van Fraassen, 2008).

Measurement involves applying some well-grounded metric, a function that assigns distances to the locations of the measured values. Applying the metric often involves setting up a scale that allows comparing the measured values. For instance, the Celsius scale allows comparing intervals between temperatures. However, not every measurement involves applying a scale and comparing measured values. Consider Neurath's *Ballung* concepts, such as race, social exclusion, well-being or quality of life. Such concepts are too multifaceted to be measured by a single metric without loss of meaning. They must be represented by a matrix of indices or several measures. Moreover, some categories are sui generis. The feeling you had the day your child was born is incomparable to any other feeling. The sublimity of the starry sky above me is not scalable.

I hold that the art of depiction is an art of measuring. Just like measurements, images are systematic mappings of the depicted state onto a representation. They are theory-laden. The acts of producing and interpreting images are embedded within our knowledge of the world. Images are selective and perspectival. They represent only selected information taken from some particular perspective.

This should not be taken as a trivial claim that images are an effect of using measuring devices such as rulers. A map is based on measurements in the same way as measuring velocity is based on measuring time and distance. In short, measurement involves measurement.

Images are not representations of the measurement outcomes, either, as if there were some measurement process separable from representing its outcomes. If a thermometer indicates 20°C, the indication does not represent some prior measurement process. It is part of the measurement.

Images are a kind of measurement devices, such as rulers (Boumans, 2005; Morrison and Morgan, 1999). Let us consider the black hole picture. It is an outcome of a measurement operation and a reliable indicator of the measured values. Just like rulers, it localizes a phenomenon in some space.

Just like rulers and balances, images represent the ways of localizing the referent. They represent the ways how to localize some states in a rule-governed manner. They are not only applications of some measures. Images exemplify these measures. Images represent the outcomes of measurements together with the rules of construction, showing how to arrive at these outcomes.

Let us consider the inflation diagram. It does not tell us only about the inflation growth value. It represents how to localize the inflation growth value by mapping time onto the inflation rate. In the same way, a world map does not tell us that London is localized at some geographic longitude and latitude only. It also informs us how to get to London.

In contrast to rulers and balances, images can represent non-numerical and qualitative properties. A colour sample shows you how to find the matching colour. A depiction of John shows how to identify John. Images can represent incomparable and unscalable phenomena. Think about the feeling you had the day your child was born. I cannot describe it or ascribe it any number. Yet, I can show you a way of how to arrive at it, for example, I can express it in music.

Thus, from a measurement-theoretic perspective, images play a twofold role. On the one hand, they are the outcomes of some measuring procedures. For instance, a portrait of John is an outcome of a measuring procedure performed on the depicted situation, in the same way as 20°C is an outcome of measuring temperature. An image localizes a depicted situation, just like 20°C identifies a temperature value. On the other, images represent measures that determine the ways we localize the measured values. Just as a ruler shows how to identify a measured length, the portrait of John shows how to identify John. Images are the tools by which we measure the world.[8]

What are the consequences of the measurement-theoretic approach for understanding the nature of images? First, the question about the nature of images is misleading. If we hold that images are a kind of measurement devices, then we can characterize images only in functional terms. Let us think about what connects mental, auditory and olfactory images. One can argue that the use of the term 'image' in all these contexts is different. That suggests, however, that there are essential properties of being an image, just like the atomic number is the essential property of being a chemical element. According to the measurement-theoretic approach, this assumption is false. Asking about the common nature of mental, auditory and olfactory images is like asking about the shared nature of rulers and balances. This is a badly posed question, for there is no essential property of being a measurement device besides bearing a measurement device function. We distinguish between different measurement instruments depending on what and how is being measured. By the same token, images can be characterized only functionally. They are the objects that are necessary for some measurement operations. Mental, auditory and olfactory images are different kinds of measurement instruments.

Second, according to the Traditional View, images are copies of the world. In the measurement-theoretic framework, they are not copies, just as the ruler is not a copy of some distance. Images may appear as if they were copies of the world, just like a ruler may appear as if it were a copy of some distance. However, just like rulers, images are measurement devices that represent the measures to localize some state of the world and help us find our way in the world.

Let us consider the portrait of John. According to the Traditional View, the portrait is a copy of John. However, we can think about the portrait not in terms of copying John but in terms of the ways of localizing John. According to the measurement-theoretic approach, the portrait is a measure that we use to identify John.

Perceiving the portrait of John does not imply that, to understand it, we need to know what John looks like; in the same way, to understand what the standard metre represents, we do not have to know what the length of one metre is. Perceiving the portrait does not imply that when we meet John, we will see him from exactly the same perspective. The portrait gives you a measure to localize John, in the same way as the standard metre is a measure to localize the length of one metre. This measure can be applied in different contexts, for it is not characterized by copying reality but searching procedures.

Compare this to a map. On the one hand, it is an outcome of some measurement procedures. On the other hand, it exemplifies a measure that helps you find your bearings in space. We do not have to know what London looks like to understand the map of London. Conversely, we use the map to learn where we are. Moreover, to find out where we are in London, we do not have to pick out the same perspective as that represented on the map. The ability to recognize its spatial invariants and apply them in different contexts is crucial for using a map.

Does that imply that the portrait of John and a map of London cannot misrepresent John and London? It does not. The portrait and the map, just like any measurement, can be inaccurate. Is this not a vicious circle, however, one could ask? To know whether a map is accurate, you need to know what it represents. To know what is represented, you need an accurate map.

This problem can be generalized and applied to the 2-D model of iconic reference. The ability to identify construction rules presupposes that we recognize the final effect of the construction. Knowing how to reach a target presupposes knowing where we want to get. At the same time, to recognize an object, we have to identify the construction parameters. For instance, to recognize the mountain seen in a painting, we must identify marks on the picture's surface that are used to represent it. However, to identify these marks, we must know that they represent the mountain. Thus, the ability to construct an object has to be taken together with the ability to recognize it.

However, the circle is not vicious. Consider the so-called coordination problem in the measurement theory. It concerns the problem of coordinating theoretical quantity terms with measuring procedures. The empirical adequacy of the theory and the reliability of the measuring procedures presuppose each other. To establish a theory, we have to test its predictions. However, this requires a reliable method of measuring. This, in turn, involves having background theoretical knowledge.

The traditional approach to the coordination problem holds that coordination is accomplished by specifying definitions for some quantity terms. These definitions are taken to be analytical and require no empirical testing. This solves the problem of circularity but forces us to accept many metaphysically loaded assumptions. The solution is particularly disturbing if we are not fans of the analytic-synthetic distinction.

According to the coherentist approach to the problem (e.g. Chang, 2004; Tal, 2013; van Fraassen, 2008), we do not have to cut the circularity of coordination. We can try to show that the circle is not vicious, for constructing a quantity concept and standardizing its measurement are co-dependent and iterative. Each iteration in the history of standardization respects and corrects existing traditions. As van Fraassen (2008) argues, the coordination problem arises when one adopts a foundationalist view and attempts to find a starting point for coordination.

Let us consider the calibration function of measurement. It is inseparable from any measurement activity, for every measurement is underdetermined by instrument indications. The same indication may be taken as evidence for multiple knowledge claims about the measured quantity, depending on which background assumptions are used to interpret the indications. Calibration is an activity of modelling different processes and testing their consequences for mutual compatibility. However, comparing some standards is neither necessary nor sufficient for successful calibration (Tal, 2017).

Let us apply the coherentist approach to the 2-D model of iconic reference. An image is an effect of a continuing and co-dependent process of searching for the best way to represent the target and trying to localize the target. It involves ongoing attempts to match a referent to a target by searching for the most precise way to identify the referent. Identifying the referent involves correcting our background assumptions regarding the target.

This process is iterative and involves modifications of image construction and representation target. It respects existing iconic conventions but is not determined by them. Most importantly, it describes both the production and interpretation of images.

Let us consider an architectural sketch. Sketching is a continuing process of searching for the most precise expression of some architectural idea. At the same time, it is a process of searching for the idea. When sketching, an architect has only a rough measure of the target. He is localizing it by producing the content.

Next, think about the process of image interpretation. It is an ongoing activity of trying to localize the target based on testing our assumptions in analysing iconic content. At the same time, it is an activity of analysing iconic content based on our background knowledge of the target. Broadly speaking, an architectural sketch and image interpretation are a kind of calibration process of modelling the idea and testing the consequences of the model.

To sum up, it seems productive to think of images as a kind of measurement devices. This solves some old problems of depiction, such as the problem of resemblance. On the other hand, it introduces some new problems, such as the one of coordination. The advantage of the measurement-theoretic approach is that it can be hoped that these problems can be solved within a more general theoretical framework.

Application: Impossible images

In the following section, I show how the measurement-theoretic account of images can be applied to impossible images. I chose this example for two reasons. First, if it is held that we can see impossibilities in pictures, then a full-fledged theory of depiction should be able to explain how to depict impossibilities. The Traditional View does not make any room for impossible images: since impossible objects cannot exist. Second, Sorensen (2002) offered a prize for a depiction of a logical impossibility. I am going to show what counts as a sensible response to his challenge.

Before going further, some restrictions are needed (Mortensen, 2010; Mortensen et al., 2013; Sorensen, 2002). First, by impossible images, I understand depictions of logical impossibilities (Elpidorou, 2016; Kulpa, 1987; Priest, 1999). Depictions of nomological impossibilities do not count, for we can have depictions of dragons and fairies. Ambiguous figures, such as the Necker cube, do not count. They entail different but consistent interpretations of logically consistent objects.

Second, by impossible images, I understand depictions of logical impossibilities open to inspection. Using a depiction of a dot as a picture of a logical impossibility seen from a long distance does not count. Inaccurately drawn geometrical figures, or figures with unnoticeable contradictory features, are not impossible images, either.

Third, by impossible images, I understand consistent depictions of inconsistent objects. Thus, if some consistent object is depicted from such an angle that it looks inconsistent, the depiction does not count as an impossible image. Similarly, we can have a visual illusion of a consistent object whose visual interpretation is inconsistent, such as the Waterfall Illusion (Crane, 1988), which is an illusion of the movement of stationary objects. Such inconsistencies are common. Yet, they do not count as impossible images.

With these clarifications in mind, I discuss two examples of impossible images: the Reutersvaard-Penrose triangle and the impossible fork. I argue that they represent two kinds of constructions. They either represent non-orientable objects or fail to construct an object.

First, let us consider the Reutersvaard-Penrose triangle (Figure 6.3).

The Reutersvaard-Penrose triangle[9] depicts an impossible object, for, in order to exist, such an object would have to violate the rules of Euclidean geometry. For example, the bottom bar of the triangle is represented as being located to the front and the back of the topmost point of the triangle. If such objects cannot

Figure 6.3 Reutersvaard-Penrose triangle. Source: Wikimedia Commons.

exist, then how is it possible to represent them? That is the paradox of impossible images.

Some philosophers and psychologists try to find a way out of this paradox by adopting three argumentation strategies. First, it is common to interpret impossible images as instantiations of perceptual illusions (Ernst, 1986; Gregory, 1966; Kulvicki, 2006a). For example, it is possible to build an object that looks like the Reutersvaard-Penrose triangle from a particular perspective.

However, the concept of an illusion implies that something merely looks like something that it is not. No such thing needs to be said about the Reutersvaard-Penrose triangle. Even if there can be a consistent object that resembles the Reutersvaard-Penrose triangle, this does not imply that the triangle represents a consistent object. If I build a scale model that looks like the Eiffel Tower, it does not imply that photographs of the Eiffel Tower represent its model.

Second, it can be held that the Reutersvaard-Penrose triangle is an example of a figure whose parts are consistent when taken separately but inconsistent when put together (e.g. Blumson, 2014; Budd, 2008; Kennedy, 1974; Voltolini, 2015). According to this argument, just as we cannot have inconsistent beliefs, we cannot have inconsistent images. Yet, we can have an inconsistent set of beliefs that can be divided into consistent parts. Priest's (1997) short story 'Sylvan's Box' is an example of such a case. According to the plot of the story, at one point, the main character believes that the box is open; at another, just the opposite. The story is an example of a more general phenomenon. Any fiction, just as experience, is full of more or less noticeable contradictions. Similarly, the Reutersvaard-Penrose triangle is inconsistent as a whole, but each part is consistent. Covering any two angles of the triangle reveals that the remaining one is consistent.

However, this analogy leaves something out. In Priest's story, we cannot simply take out parts of Sylvan's beliefs and hold that they are consistent. Such a move would turn any contradictory system into a non-contradictory one. We hold that the system is inconsistent if the relations between the parts of the system are inconsistent. Thus, we need to supplement the story by holding that Sylvan can be unaware of the inconsistencies of his beliefs. Alternatively, he can have some implicit beliefs that render the inconsistent story consistent. No such thing applies to the Reutersvaard-Penrose triangle. It is neither a depiction of three disassembled parts nor is there anything implicitly assumed in the picture. We see three assembled parts that are together inconsistent. Nothing is hidden or implicitly assumed. The impossible object is seen in the picture.

Third, the moral of Priest's story can be interpreted differently. Instead of asking about the consistency of the story's parts, we may ask what can render this story consistent. In 'Sylvan's Box', the storyteller is reporting Sylvan's beliefs that the box is empty and not empty. Priest does not believe that it is possible. But it can be possible *within* the story in the same way as we can affirm contradictions in some logical systems, such as paraconsistent ones.

The same can be said about the Reutersvaard-Penrose triangle. It is a consistent representation of some consistent object in some non-Euclidean space that looks like a three-dimensional Euclidean space (Francis, 2007). It looks like an inconsistent representation, for we implicitly assume that we deal with an ordinary Euclidean space, which is false. Alternatively, we can stay within Euclidean geometry and interpret the triangle as a type of the so-called occlusion paradox, which means that it can be rendered consistent by changing one or more occlusions (Mortensen, 2010, 119). You get a consistent triangle representation if you rotate any two corners of the triangle.

Discussing the details of both solutions will not get us far, however, for they cannot tell the whole story. Most of all, they leave us with a gap in our understanding of the following example of an impossible image. Let us consider the impossible fork (Figure 6.4).

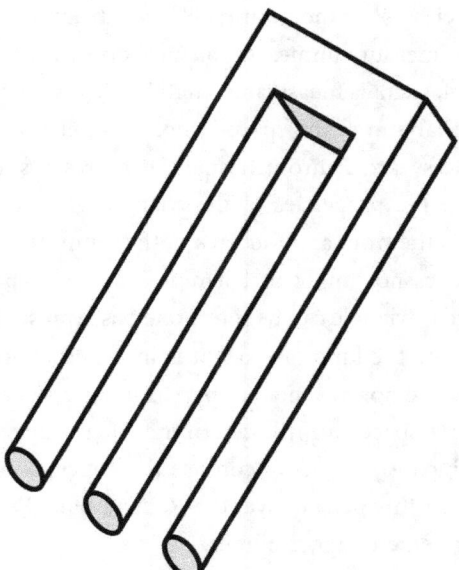

Figure 6.4 The impossible fork. Source: Wikimedia Commons.

The impossible fork depicts an object that must have two and three prongs simultaneously to exist. This cannot be the case. Therefore, it is an impossible image.

No consistent mathematical theory can describe the impossible fork (Mortensen, 2010, 135 ff.). It is not a type of an occlusion paradox, either. Changing the occlusion of any part(s) of the figure does not make it consistent. According to Mortensen, the only way out is to acknowledge the existence of paraconsistent geometrical systems that can include objects such as impossible forks.

However, the price is that we have to acknowledge the existence of contradictory objects in our ontological landscapes. Consequently, we have to adopt some sort of dialetheism, which is a belief that there are true contradictions (Priest, 1999), which is not a conclusion everyone can happily accept.

Dialetheism might or might not be true, but we can do without it. According to the measurement-theoretic account, images refer to their objects by exemplifying how these objects can be constructed. The general idea is that there can be no impossible objects, but there can be 'indeterminable' constructions.

Constructions have been characterized as procedures of arriving at a target by means of determining the parameters of some space. Arriving at a target means localizing an object in this space. However, there are such constructions whose parameters cannot determine the properties of some space to localize the target. These are indeterminable constructions.[10]

There are two classes of indeterminable constructions. We can compare them to attempts to measure unmeasurable properties and attempts to measure with the wrong tool, such as measuring length with gas. The first class refers to the constructions that aim at the spaces whose properties are indeterminable. Consider the Reutersvaard-Penrose triangle. It represents a construction that does not determine the properties of the triangle, for the bottom bar of the triangle is located to the front and the back of the topmost point of the triangle.

However, this does not imply that it represents an impossible object. The Reutersvaard-Penrose triangle can be interpreted as representing a construction that cannot determine the kind of space it is in. Alternatively, it can represent a construction in some non-orientable space, for example, a four-loop Möbius strip. Non-orientable spaces do not determine such properties as orientation. They may appear inconsistent if we confuse the concepts of non-orientability and inconsistency. In this perspective, the Reutersvaard-Penrose triangle only appears as if it represented impossibilities.

The second class of indeterminable constructions refers to the constructions that fail to determine the properties of some space. These are abortive constructions. They cannot reach a target since the construction parameters are such that they localize an inconsistent object. Recall the operation of dividing by 0. It localizes a set of inconsistent values. Constructing a triangle from the line segments such that each side of the triangle is bigger than the sum of the other two sides is abortive, for the operation cannot construct a triangle. However, the fact that some constructions are abortive does not imply that they do not exist.

The proper class of impossible images refers to such images that exemplify abortive constructions. The impossible fork represents the rules of construction that cannot reach the target. It is only one of the many abortive constructions.[11]

Moreover, since we distinguish between the properties of objects and the properties of construction, we can keep the intuition that impossible objects cannot exist, but there can be representations of impossible objects. Impossible images represent constructions without targets. Consider a round square. Can it exist? Obviously not. But there can be an image of a round square. Why is this so?

Metaphysical impossibilities are logical contradictions. The concept of a round square is clearly contradictory. If so, according to the Duns Scotus Law, if p & not-p, then q, everything can be an instantiation of this concept. However, this does not seem right, for it contradicts the intuition that there can be better or worse representations of non-existent objects. Compare an image of a shape representing a cross between a square and a circle (the so-called squircle) and an image of a horse. Granted, they are both inaccurate depictions of a round square. However, the squircle appears less inaccurate than a horse picture. The squircle represents an inconsistent measure to localize the round square, but it is still better than the horse measure. By the same token, a ruler made of elastic material is a bad measure, but it is still better than a ruler made of gas. In both cases, we will get inconsistent results; yet in the first case, they will be less inconsistent.

To sum up, in the measurement-theoretic framework, we do not have to introduce inconsistent geometries to explain impossible images. Moreover, we can deliver a unificatory explanation of two kinds of impossible images. Impossible images are the kind of images that exemplify indeterminable constructions.

Summary

In this chapter, I described the 2-D model of iconic reference. According to the model, images have two-dimensional semantics. They denote their targets and exemplify content that identifies the referent. In the 2-D model, images employ two kinds of correctness conditions characterized in the measurement-theoretic terms: trueness and precision. Images that are both true and precise are accurate.

I employed the measurement-theoretic framework to explain the properties of iconic reference. I claimed that images are a kind of measurement devices. Contrary to the Traditional View, images are not copies of the world. They are ways to localize the properties of the world, comparable to rulers.

In the last section, I showed how this framework could be applied to explain the phenomenon of impossible images. According to the measurement-theoretic view, impossible images represent indeterminable constructions.

7
Thinking with images

In the last chapter, I described images as a kind of measurement devices. I argued that images represent the world in a twofold manner: by denoting their targets and exemplifying the rules of construction that reidentify the referent. If the referent matches the target, then an image is accurate. In contrast to the Traditional View, images are not copies of the world. They are measurement devices, such as rulers and balances, used to localize the states of the world.

In the following chapter, I demonstrate how the measurement-theory account of images can help explicate the imagistic theory of thought. Primarily, I explain how this account addresses three kinds of challenges to the imagistic theory of thought. According to the Epistemological Challenge, the imagistic theory of thought cannot provide a theory of knowledge, for images are not truth-evaluable and lack logical form. According to the Semantical Challenge, the imagistic theory of thought cannot provide a theory of content as the object of imagistic thought is impossible to determine. According to the Metaphysical Challenge, the imagistic theory of thought cannot explain the systematicity and compositionality of thoughts.

Finally, I show how the measurement-theoretic account can be applied to the problems of representational format and mental imagery. Particularly, I demonstrate the metaphysical consequences for thinking about the structure of mental representations and the nature of mental imagery.

Imagistic knowledge

The backbone of the Epistemological Challenge is Wittgenstein's argument from content indeterminacy and Frege-Davidson's argument from lack of logical form. Their common assumption is that they implicitly posit the Traditional View. However, these arguments do not apply to the measurement-theoretic account of images. Why is that so?

The general idea of the Traditional View is that images resemble the world. From such a perspective, it is always possible to ask whether iconic content fits the resembled reality. Thinking about imagistic content in terms of matching the world suggests that images can be true or false. However, these assumptions are questionable if we accept the conclusions of Wittgenstein's and Frege-Davidson's arguments. There is no room for pictorial truth.

It may appear that the lack of truth values prevents images from being part of a rational train of thought. However, we can accept the premise without acknowledging the conclusion. We do not need truth for something to be part of a rational train of thought. Models and idealizations are not true, yet they are a vital part of knowledge (Cartwright, 1983; Elgin, 2017). A thermometer is not a truth-bearer. However, its readings can be part of justified beliefs.

In the same way, images can do epistemic work without being true (in the propositional sense). A geometric drawing justifies our beliefs that a mathematical figure is constructible. A map justifies our belief that we can get from London to New York (e.g. Gauker, 2020; Shepard and Cooper, 1982; Williamson, 2016).

Moreover, images can test our beliefs. If I believe that triangle sides can have lengths such that each side is longer than the sum of the other two, I can change my mind having been shown that such triangles cannot be constructed. If I hold that London is west of New York, I can change my mind by looking at a map.

Does this mean that images provide a reliable source of knowledge? It depends on our epistemic goals. Using measurement devices is context-sensitive. Rulers are usually reliable measurement devices, but sometimes they fail to meet our needs. A ruler can identify the length of line segments, yet it is useless when it comes to measuring the distance between the Earth and the Sun. Usually, we rely on images. Sometimes they are unreliable. However, this does not imply that we can dismiss images as useless in our epistemic practice, just like we do not dismiss rulers as useless.

Thus, images appear to provide some knowledge. The question is what such imagistic knowledge can be. If images are not true, they cannot provide propositional knowledge, for the latter involves a truth-evaluable description of the world. By the same token, images cannot provide information about possible states of affairs, for it would require them to be truth-bearers. Neither can they provide explanations, for the latter are propositional and have logical form. The explanation is shaped by argument and involves inferential structures. Images are not 'inferentially promiscuous'.

It may seem that we can respond to these doubts by biting the bullet and distinguishing between propositional and non-propositional knowledge. One can hold that although images are not propositional, they can constitute a form of non-propositional knowledge.

However, much depends on how the concept of non-propositional knowledge is understood. It is most often argued that imagistic knowledge, contrary to propositional knowledge, is non-conceptual in content and builds phenomenal knowledge or knowledge-by-acquaintance (e.g. Gauker, 2011; Gregory, 2013; Mößner, 2018). For instance, a memory image of a red rose gives us knowledge of what rose-redness looks like (e.g. Tye, 2012).

This strategy, however, will not get us far. First, non-propositional knowledge can have the same correctness conditions as propositional knowledge. Depicting what a thief looks like can be put in terms of a set of propositions attributing properties to the thief. For instance, depicting a thief as bald is correct under the same conditions as my belief that the thief is bald. If I depict the thief as having long hair, I depict him incorrectly in the same way as the proposition THE THIEF HAS LONG HAIR may be incorrect. That implies that if I know what the thief looks like, I know that he is bald. In other words, if I know what x looks like, I know that in such and such circumstances, x looks so and so (e.g. Stanley and Williamson, 2001).

Second, we rightly expect images to inform us about the world. For instance, if I take a picture of a thief, I expect to learn who the thief is, whether he is bald or tall. I am not interested in how the thief is presented to my mind. When scientists study the black hole picture, they do not want to learn about the phenomenology of black holes, either. They want to know what black holes are.

Thus, images should tell us how the world is and not only what it looks like. If images could provide only phenomenal knowledge, then imagistic knowledge would be useless from the point of view of our epistemic interests. Instead, we rightly expect images to figure within the web of our beliefs and theories.

The Epistemological Challenge questions these expectations. In other words, the role we attribute to images cannot be fulfilled. That is because we hold that beliefs and theories are truth-evaluable; images are not. Images are not propositional, yet we expect them to figure in propositional knowledge. The problem is not that these expectations are not right. The question is how they should be explained.

The lack of propositional knowledge can be compensated by an image's contribution to understanding. As it seems, images are a plausible source of understanding without being a tool of explanation (e.g. de Regt, 2017; Lipton, 2009, Mößner, 2018; Zagzebski, 2019). The black hole picture gives us an insight into what black holes are. It does not provide a theoretical explanation of why they are so. That being said, we need a clear account of what understanding is.

Importantly, we need a view of understanding that is not reducible to explanation. For instance, coherence-based accounts of understanding (e.g. Elgin, 2017; Kosso, 2007; Kvanvig, 2018) define it as recognizing connections between facts and as a skill of matching things to form a general schema. However, recognizing coherence between facts is only another way of knowing an explanation of these facts since the explanation is based on identifying factual connections (Khalifa, 2017).

According to de Regt (2017), understanding is distinct from explanation. It does not require true theories, either. It is a skill of making theories intelligible but not necessarily true. A theory is intelligible if scientists can recognize some characteristic consequences of the theory qualitatively. For instance, a kinetic picture of gas as a collection of molecules in motion enables a qualitative understanding of how gas behaves. By developing molecular models of gas, one can arrive at intelligible kinetic theories of gas which are the sources of constructing explanations of the gas phenomenon.

Although de Regt's concept of understanding does not directly apply to imagistic knowledge, it hints at how to think about the latter. According to the measurement-theoretic account, images are measurement devices representing the ways of identifying measured objects. The idea is that they enhance understanding by providing recognition-based identification of the objects of our beliefs and theories.

Recognizing objects is not a propositional kind of knowledge. It is not a matter of knowing a description of an object, either. I can know a description of an object without being able to recognize it. Recognition is the skill of localizing the represented object in some space. Knowing how to recognize an object is a non-propositional (procedural) kind of knowledge.

Let us compare this to the understanding of indexicals. According to Perry (1979), understanding indexicals such as 'here' and 'now' is irreducible to knowing a description of the space you are in. Believing that a meeting starts at noon is irreducible to knowing that it is now. Knowing that to leave a wilderness, you have to follow the Mt Tallac trail is different from knowing that the Mt Tallac trail is here. Indexicals locate our beliefs in space and time.

By the same token, imagistic knowledge provides an understanding of the objects of our beliefs and theories by localizing these objects in some logical or physical space. What does it mean?

Imagistic Knowledge: the skill of recognizing objects of our beliefs and theories within the logical and physical space.

Let me illustrate this with the case of interpreting a map (van Fraassen, 2008). To understand our position in space, we have to localize ourselves on the map. This is not the kind of information we can get from the map's description. Unless we can find our location on the map, the map is useless. Even if a map indicates the point 'we are here', we still have to determine our position. Based on the map, we have to localize ourselves by identifying the parameters of the space. Moreover, we need to know what moves are permissible by the map. For instance, we need to know that if we turn left, we can get to point A but not B. To interpret the map, we must identify how we can move.

The operation can be quite simple. It can be based on reading off the information 'Diagon Alley' and finding the same inscription on the street. Next, based on our knowledge of how streets are plotted on the map, we can find our way from Diagon Alley to Knockturn Alley. Thus, a successful interpretation of the map involves our ability to localize and orient ourselves in the space. What does this teach us about the nature of imagistic knowledge?

According to the 2-D model, a map is a measurement device used to orient ourselves in the logical or physical space. It exemplifies the rules of construction and construction invariants. These invariants are the landmarks, such as street names and the spatial relations between them. They enable us to identify spatial localizations and directions. A successful interpretation of the map involves identifying these spatial invariants and recognizing permissible moves so that one can orient oneself in space. Consequently, a good map helps us to understand our position in the physical or logical space.

By the same token, imagistic knowledge enhances the understanding of our beliefs and theories by means of localizing them in the logical or physical space. Such knowledge is irreducible to the set of propositions in the same way as understanding a map is irreducible to knowing the map's description. Imagistic knowledge is the skill of recognizing the objects of our beliefs and theories within the experience and the space of possible states. Let me illustrate this with two examples.

Suppose you want to understand what a harmonic oscillator is. This involves being able to point at a pendulum and hold 'this is a harmonic oscillator'. The pendulum is a model that identifies the invariant properties of the harmonic oscillator by means of which you localize the harmonic oscillator within the space set up by classical mechanics. The model does not explain the theory. It localizes the states set up by the theory. It helps us to orient ourselves within the theory, making it intelligible.

In the same fashion, the black hole picture makes the theory of general relativity intelligible.[1] The theory offers explanations of the origin and behaviour of black holes. It provides a theory of what black holes are. To understand this theory, you have to be able to localize the object of the theory in the physical or logical space. You have to be able to point at the black hole picture and hold 'that is a black hole'. The black hole picture enables recognizing the phenomenon the theory describes. It helps you to orient yourself within the theory of general relativity.

Knowing how to orient ourselves within the theory is a non-propositional kind of knowledge. It consists in knowing how to localize the object within some logical or physical space set up by the theory. Such information cannot be deduced from the theory. The role of a theory is to explain a given phenomenon. In contrast, images localize the object of the theory by enabling recognition-based identification of this object.

By the same token, images identify the objects of our beliefs. Let us consider a portrait of John. The picture identifies John, just as indexicals 'here' and 'now' localize an object in spacetime. The portrait exemplifies the construction rules by means of which we identify John. Such information is irreducible to knowing a description of John. I can recognize John without knowing a description of John; I can know a description of John without any ability to recognize him in the picture, too.

Images are irreplaceable in imagistic knowledge. Knowing how to recognize objects is irreducible to knowing the descriptions of these objects and, therefore, cannot be represented by any set of propositions. The picture of John can misrepresent John, but it does not express any false proposition. I can recognize John in the picture without any descriptive knowledge of who he is. The only known medium that can represent imagistic knowledge is an image, for it exemplifies the rules of construction by means of which we identify the depicted object.

To illustrate this irreplaceable role of images, let us consider the mental rotation tasks (Shepard and Cooper, 1982; Shepard and Metzler, 1971). The task is to create a mental image of an object and rotate it in order to compare it to the presented figure. Next, one has to decide whether both objects are the same.

Notice that such a decision cannot be made simply by comparing descriptions of the object expressed in some language-like representation system (Pylyshyn, 1973). First, you need the skill of rotating the first object in the right way. There are infinitely many ways to rotate this object, but only one way leads to solving the problem. The way of rotation cannot be inferred from the figure's description. If it could, it would require checking infinitely many ways of rotation which leads directly to combinatorial explosion.

Second, we would still need additional information if we had two similar descriptions of the figure. The task is to identify two objects as identical. This information does not follow from having two similar descriptions. I can know these descriptions without any ability to recognize that they refer to the same object. I can know both the content of a map and a description of my environment, but it does not follow that I can orient myself in the space. By the same token, I can know a description of John, but I still need the information that the depicted person is John (Kaplan, 1989).

Granted, the description-based explanation of how people solve mental rotation tasks does not have to be wrong. The problem is that it is not sufficiently informative to work as an explanation.

In the 2-D model, the way we solve mental rotation tasks is easier to understand. Recognizing two objects as the same involves identifying the rules of construction. Images exemplify the rules of construction by means of which we recognize depicted objects. We recognize two figures as the same if we identify the same construction invariants determined by these rules.

In the mental rotation tasks, subjects hold that they visualize the process of the figure rotation. Suppose the measurement-theoretic account of images is correct. In that case, the fact of visualizing this process results directly from the epistemological properties of images. Only images can convey the information that is needed in mental rotation tasks. The pictorial character of mental imagery is not a contingent fact of human psychology but a necessary semantic fact.

Although imagistic knowledge is irreducible to propositional knowledge, they are both entangled. First, localizing the phenomenon that is the subject of a theory enables testing the theory. The epistemic role of images is to localize the states within the space set up by the theory. If these states contradict each other, it is a good reason for rejecting the theory.

Let us suppose that the black hole picture contradicts our predictions resulting from the theory of general relativity. If a theory predicted that black holes were square-shaped, but the black hole picture identified a crescent shape, it would be a good reason to hold that the theory is wrong rather than poorly illustrated by the picture.

Second, images can be a basis of propositional knowledge without having a propositional character. According to the 2-D model, images do not describe the world. They represent the ways we identify the states of the world. Images represent how the world can be perceived. Consequently, images are not true or false in the way propositions are. Images can accurately or inaccurately localize the target. They are not expressing any truth. Propositions do that.

However, based on identified objects and properties, we can form beliefs and theories. Images are the measurement devices we use to inform ourselves about the world. A measurement for its own sake is a fruitless endeavour.

Let us consider the case of a ruler. The ruler's length is neither true nor false. It represents the way we isolate a spatial magnitude. At the same time, the indication of the ruler can provide a basis for describing distance. Based on that indication, I can hold that a line segment is one metre long and so on. We use a ruler to describe its target.

Similarly, we use images to describe the world. Pointing at the black hole picture, I can hold that the black hole emission ring is crescent-shaped. The picture is used to describe the black hole by isolating its spatial features.

Images enable recognition-based identification of objects. However, images cannot tell us what these objects are. To interpret an image, we need propositional knowledge.

Compare this to the way we interpret measurement indications. There is a difference between measurement indications, such as the numerals on the thermometer display, and measurement outcomes. The latter are knowledge claims that associate the measured values with the object being measured and are inferred from one or more indications along with relevant background knowledge (Tal, 2017). The reading of the thermometer indicates some physical value. However, it is the role of a theory to interpret this value. Without a theory, we can identify objects but cannot know what they are.

Similarly, using images involves some background knowledge necessary to interpret their meaning. For instance, the black hole picture provides perceptual recognition of the black hole. Yet, we need a physical theory to know that the picture represents a black hole.

With these distinctions in mind, we are now in a position to address the problem of phenomenological content. Let us recall Gregory's idea of distinctively sensory content. According to Gregory (2013), pictures represent the ways something is like. They represent qualitative aspects of experience. For instance, Cezanne's Mont-Saint-Victoire paintings represent distinctive ways in which the mountain appears to us. They represent the colour depth and the way the light illuminates the mountain. How should we explain this?

The Traditional View appears to be unable to explain the phenomenological nature of imagistic knowledge. First, the Traditional View requires comparing two relata of the resemblance relation. This implies that we would need to access two spatiotemporally different quale simultaneously, which cannot be right. I cannot have simultaneous access to how Cezanne's Mont-Saint-Victoire paintings look and what it is like to see the mountain.

Second, it gives us no insight into how to determine the conditions of correctness, for it would require access to the painter's private experience. Cezanne painted several versions of Mont-Saint-Victoire. Without knowing what his experience was like, we cannot determine which of the paintings represents the mountain correctly.

From the measurement-theoretic perspective, to understand the sensory content of a painting, we need to be able to localize the represented quale in some space. Let us think of it in terms of constructing a map. Suppose that every quale represents a point on this map. Image properties determine the parameters of the space that the map represents. They are the means by which we localize a particular point on this map. Images are measurement devices we use to identify this point.

Consider a particular shade of green represented in Cezanne's painting. According to the 2-D model, it is not a copy of the quale experienced by Cezanne. It is the construction parameter by which we identify the particular shade of colour. The painting represents the way we localize this colour in our experience. Cezanne's painting exemplifies a measure we use to recognize this colour in experience. For instance, we can point at some colour and hold 'that is the colour I have seen in Cezanne's painting'. We see through Cezanne's paintings just like we identify length by using a ruler.

Importantly, the ability to identify a particular shade of the colour in the painting is necessary to understand this colour. Consider two cases. First, suppose that you know a description of the colour in the sense of knowing its coordinates in the RGB system. To make this description intelligible, you need to be able to localize it in some logical and physical space. Primarily, you need to know what RGB stands for, and therefore you need to know, in advance, what red, green and blue are. In other words, you need the skill of identifying these colours.

Second, we can describe a colour shade as a particular wavelength of light experienced by a subject. However, this wavelength of light is precisely this shade of colour (Williamson, 2002). The ability to find this colour is part of understanding this description. A colour sample in the painting does exactly that. It localizes this description within the space of colours.

According to the 2-D model, Cezanne's painting is a measurement device by which we identify the world's phenomenological properties. This does not imply that it provides information on what these properties are. There is always some background knowledge involved in the process of interpretation. For instance, we have to know some facts about the author to know that the painting represents Mont-Saint-Victoire. That is not the kind of information that we can get from the painting. However, this is not the image's purpose.

In a nutshell, images exemplify measures by which we identify properties, objects and events. They enable us to understand the objects of our beliefs and theories by means of localizing these objects in some logical and physical space.

Iconic content

According to the Semantical Challenge, the imagistic theory of thought cannot provide the theory of content. Images cannot determine the content of the predicative function and individuate the object they predicate of.

The Semantical Challenge and the Traditional View are closely bound together. According to the Traditional View, image A refers to object B in a way that A resembles B. However, to determine that A resembles B, we have to individuate the object that is being resembled. To individuate this object, we have to determine what properties are depicted. Thus, the structure of resemblance relation mirrors the structure of a predicative function.

The Semantical Challenge holds that predication requires a structure that distinguishes between the object of predication and the property we predicate. Propositions and language have such a structure. Images do not, for their content is indeterminable, and they lack logical form.

In the 2-D model, the Semantical Challenge is misdirected. Primarily, the content of images does not involve any predication. Iconic content is not a description of some object. It represents the way of localizing this object.

Let us consider thermometer readings. If I hold that the air temperature is 20°C, I do not hold that the temperature is in some dyadic relation to the number 20. I do not hold that 20 resembles the temperature, either. The number 20 does not predicate of this temperature that it has the property of being even or being twice as much as the temperature 10°C. The number 20 is a measurement predicate that represents the way we identify a physical magnitude by finding its place on a scale. This number localizes the particular state of the world and its relation to other states. It shows how this temperature can be localized in some measurement set-up.

By the same token, an image of John being bald represents the way of identifying John. Depicted properties are the information we use to localize John. When we depict John as bald, the property of being bald works as the measurement predicate used to identify John. It does not predicate anything of John, as if there were some relation between these properties and their object seen in the picture, that is, something we predicate of and a predication. Depicted properties represent the way we see John, not the way John is.

Let us compare a depiction of John with the proposition JOHN IS BALD. The proposition individuates the object of predication, that is, John, and predicates of him the property of being bald. The property of being bald is the property of the predicated object.

In contrast, we do not hold that the marks on the picture's surface are bald or that John is made of paint. These marks are the properties of the representation, not the properties of the represented object. According to the 2-D model, the properties of representation are the properties of the construction. Thus, representing John as bald identifies the ways of arriving at John. Mistaking the properties of construction and the constructed object leads directly to the pictorial fallacy.

In the measurement-theoretic perspective, images do not bear predicative functions. However, images can be a basis of predication in the following way. Suppose you measure some object with a ruler. The indication of the ruler, such as 'one metre', does not predicate the property of being one metre of this object. It identifies this property. By using a ruler you establish the length of the object. However, once this value is found, we can predicate it of this object by the proposition THIS OBJECT IS ONE METER LONG.

By the same token, a portrait of John does not predicate anything of John. It localizes John by determining the properties that serve to recognize him. For instance, it represents the property of being bald which is a measurement predicate used to identify John. Once you recognize John in the picture, you can predicate of John that he is bald.

Language is essential to express predicative functions (e.g. Davidson, 2001; Devitt, 2006). It distinguishes between the object of predication and the predicate. It distinguishes between different modal contexts. Images cannot do that. However, their job is to identify objects, not to predicate of them.

Moreover, the role of language is to determine the kind of predicate we project onto the predicated object. Consider an image of a red square. It identifies the properties of being red and square. However, the image does not predicate of red that it is square-like or of a square that it is red. Language does that. Only within the linguistic categories can we distinguish between the meaningful statement that a square is red and a meaningless expression that red is square-like.

In the measurement-theoretic account, language and images are semantically linked. On the one hand, recognizing the iconic content is a prerequisite to describing this content in language. For instance, if pointing at a picture of John, I hold 'it's John', I have to be able to recognize John in the picture. Recognizing John is a skill that is not language-based. In contrast, it is necessary to form a description of John.

On the other hand, language is essential for making images meaningful. I can recognize John in the picture, but without language, I cannot represent the thought that the picture represents John. By the same token, we need language to interpret the measurement's results. Thermometer readings are meaningless unless we can express what they represent. Only language can do that.

Does this imply that the iconic content is indeterminable and needs language-like representations to fix its meaning? It depends on how we understand this question.

Let us make two preliminary remarks. First, to determine the content of representation, we need a criterion by means of which we decide what the object of representation is. Second, in the 2-D model, we distinguish between rigid and non-rigid kinds of reference. Respectively to the context of evaluation, an image denotes an object non-rigidly. The iconic content, however, works as a rigid designator. It identifies the same referent in every possible world.

The general idea is that we can introduce the criterion only in the case of non-rigid kinds of reference. It is always possible to ask about a criterion that determines image denotation. For instance, I can hold that the target of the portrait of John is John since the author intended to depict John. Here, the intention is a criterion by means of which we identify the target respectively to the context of evaluation. We use this criterion to determine the target of the portrait. Moreover, propositional representations are essential here. Applying a criterion requires that we have a set of beliefs that are a basis of background knowledge that justifies the use of such a criterion.

The question about criterion, however, is senseless in the case of iconic content. We cannot ask about the criterion that determines iconic content. This type of content refers rigidly to the referent. It identifies the same referent in every possible world regardless of the context of evaluation. Iconic content is criterion-less and directly referential. Let me explain.

According to the 2-D model, images exemplify the rules of construction by means of which we identify the referent. Here, the question about criteria cannot be brought up. Just like a ruler sets up the criteria for identifying distance, an image sets up the criteria of referent identification.

Let us compare this to the Traditional View. If we say that pictorial content resembles some object, we can always ask about the criterion of resemblance. It is possible since we can distinguish between the copy and the copied object. Moreover, the resemblance criterion appears to be primitive and basic. It does not involve any dictionary or a syntactical structure. We directly see the resemblance. Both assumptions, however, make the Traditional View vulnerable to Wittgenstein's indeterminacy argument.

In contrast, in the 2-D model, iconic content is not a copy of the target. It is a representation of the ways by which the target is identified. Here, the question about a criterion is meaningless, for we cannot distinguish between what is being represented and how it is being represented. There is no place for a middle term between the image and a referent, for the image exemplifies the referent.

The source of the confusion is that iconic content represents some rule-governed procedures of arriving at some object but is not a description of this object. Images exemplify the way this object can be taken.

To illustrate this, let us consider the following example. Suppose that I show you how to construct a triangle. If I am drawing a triangle, I am not predicating anything of this triangle. I am showing you how to arrive at it. This drawing has certain properties, for example, the lengths of the line segments. These are the properties of construction, allowing you to identify a triangle. The lengths of the line segments are the properties of the searching procedure by means of which you arrive at the desired object.

At the same time, the length of these line segments identifies the object's property. If I show you how to construct a triangle, I arrive at some object with determined side lengths. The length of the triangle line segments is the final effect of some construction procedures. We cannot separate these properties and ask about the criterion by which we determine representational content. The way the object can be depicted *is* the content of the depiction. The image demonstrates that if you take these line segments and connect them properly, you will arrive at this object.

Importantly, we should not confuse the properties of construction with the properties of the constructed object. The triangle construction identifies the length of the searched object's line segments, yet, it does not describe this object. By the same token, the length of the ruler represents the way of identifying the measured magnitude; yet, the length of the ruler is not the property of the measured length. The ruler represents the way of localizing the object's length.

Let us compare this with a description of a triangle. If I describe what a triangle is, I predicate some properties of the mathematical object. The description can be wrong, depending on whether the properties identified by the description match the object's properties. You need to find the described object to determine whether a description is correct.

In contrast, a triangle diagram does not have to be compared to some abstract mathematical object. The image tells you that if you take such and such lengths of the line segments and arrange them properly, you will arrive at the object represented by the image. With a triangle diagram, you do not say that now you can look for triangles to find out what they are. You do not have to look for anything that can match the image. The image is the thing you are looking for.

Compare this to the portrait of John. Suppose that you want to know who John is. In this case, I can point at his portrait. The portrait represents a way of identifying John. However, if you are shown a portrait of John, you do not say that now you know how to find out who John is. The portrait is the thing you are looking for. The portrait exemplifies the standard you use to identify the referent. It does not require a criterion you use to identify the content, for its iconic content determines the criterion of referent identification. The portrait of John exemplifies the criterion by which you identify John, in the same way as a triangle diagram exemplifies the criterion by which you identify triangles.

By the same token, a ruler exemplifies a standard used to identify spatial distances. I can point at the ruler's length if you ask me what one metre is. However, it is nonsensical to say that now you know how to find out what one metre is. The ruler's length is not a description of some spatial magnitude. It sets a standard to isolate some spatial magnitude. You cannot ask for a criterion to determine the correctness of this standard, just as it makes no sense to ask how you know that the standard metre is one metre long.

Importantly, this does not imply that these standards cannot be inaccurate. The triangle diagram can be badly drawn; the portrait can be imprecise. Depending on the context of evaluation, the referent can match no target and so on. However, the question of accuracy conditions of some representational standards should be strictly distinguished from the question of how we know what these standards represent.

Let us apply these considerations to Wittgenstein's example of a picture of an old man walking up a steep path that looks as if the old man is walking down. Two questions can be raised here. We can meaningfully ask whether the image accurately identifies the target. In the same way, we can meaningfully ask whether the indication of the ruler represents the distance or the velocity in some frame of reference in relativistic physics.

However, it is meaningless to ask how we know that the iconic content is the man walking up, not the man walking down. The iconic content permits both interpretations, for it exemplifies both construction rules. These rules determine the standards of interpretation. There can be no question of how I know which rule it exemplifies, for we directly see the rule in the picture.

Think about how we interpret the content of the Necker cube. Depending on the applied construction rule, it can be interpreted differently. However, this does not imply that its content is indeterminable but only that it can be interpreted differently. Moreover, it is meaningless to ask how we know that we see one interpretation of the Necker cube and not another if the shape of the Necker cube remains the same. The only answer is that we see it as such. The construction rule we identify in picture-perception determines the way we interpret the picture.

By the same token, Wittgenstein's picture of an old man can be interpreted differently, depending on the identified construction rules. However, there can be no question about the criterion we use to identify the construction rule since construction rules determine the interpretation standards.

Iconic content is directly referential. We cannot ask how we know that we recognize an old man as walking up or down. We see it as such. By identifying the construction rules, we directly see something in the picture. This is the reason why it may appear that iconic content is primitive and unanalysable.

Thus, in the 2-D model, Wittgenstein's argument appears to miss the mark. Iconic content is indeterminable in the sense that it is criterion-less and directly referential. This does not imply that iconic content is more primitive or basic than propositional content. The rules of construction must be identified in the picture to recognize an object. Such identification requires skill and practice.

Moreover, holding that iconic content determines the standard of object identification does not imply that the standard cannot be wrong. It can be inaccurate, depending on what and how it is being represented.

The criterion-less and direct character of iconic content means that the latter sets up the criterion for recognizing objects, properties and events. Iconic content does not require any criterion of interpretation, for it determines this criterion. Images represent how the world can be perceived in the same way as rulers show how objects can look in some measurement set-up.

Metaphysical constraints

The Metaphysical Challenge concerns the problem of systematicity and compositionality of thoughts. According to the challenge, images are not systematic and lack canonical decomposition. Therefore, they cannot be bearers of thoughts.

The backbone of the Metaphysical Challenge is the belief that images have no logical structure, for the division of images into representational parts is arbitrary (Casati and Varzi, 1999).[2] Let us recall Fodor's Picture Principle. It holds that if *P* is a picture of *X*, then parts of *P* are pictures of parts of *X*. According to Fodor (2008), all parts of a picture are its constituents. Take a picture of a scene and cut it into arbitrarily chosen parts. Every picture part will represent a part of the scene. Consequently, iconic representations do not have a canonical decomposition, that is, a correct way to determine syntax and lexical primitives.

Fodor's argument is based on the premise that every image-part is representational. However, this premise is clearly false. Canvas texture is part of a picture, but it does not represent the scene's texture. The distances between points are parts of a topological map, yet they do not represent the distances of the mapped scene.[3] What seems to be the problem?

First, Fodor operates with an ontologically restricted understanding of the part concept. He takes it as referring to some events, material bodies as well as spatiotemporal and geometrical regions. However, the concept of part is ontologically neutral. It can refer to abstract entities such as properties, propositions, types and kinds, too. For instance, one can hold that extraversion is part of personality. Fodor's premise that iconic representations can be divided arbitrarily appears untenable with the ontologically neutral understanding of the part concept. Take a gustatory image. It can be divided into property classes such as salty, sweet, strong, minty but not in any way we choose.

Second, the Picture Principle premise appears acceptable only if we adopt an unsophisticated version of the Traditional View, according to which pictorial parts are pixel-like primitives. However, pixels are primitive when it comes to information registration but not when it comes to representational content (Burge, 2018). Analogously, phonemes are not representationally primitive from the perspective of the sentence meaning; yet phonemes have to be registered to form a sentence.

Luckily, the Traditional View is not the only game in town. According to the 2-D model, iconic content consists of the construction rules by means of which we identify the referent. Constructions have been defined as operations determining the parameters of some logical or physical space. The idea is that we can distinguish between representational and nonrepresentational image parts by identifying construction parameters.

Let us consider the triangle diagram. It represents the way a triangle can be constructed. The triangle construction highlights the properties such as the number of lines and the angle measure. These are the properties highlighted by the construction parameters and content constituents. In contrast, properties such as the colour of the lines are not representational as they are not singled out by the construction parameters.

Image parts can be representational or nonrepresentational depending on the identified construction rules. If I want to construct a congruent triangle, the size of the lines is representational. If I want to construct a similar triangle, the shape of the triangle is what matters. The size of the lines is nonrepresentational.

Let us compare this with the black hole picture. The construction rules identify the crescent shape of the black hole emission ring. The properties of the shape are representational because they are highlighted by the construction parameters. In contrast, the photograph's texture is nonrepresentational, for there is no explicit construction rule in which the texture is highlighted by any of its parameters.

Let me underline this point. In the 2-D model, there are no metaphysically or psychologically predetermined representational primitives, such as lines and dots (e.g. Burge, 2010, 2018; Camp, 2007; Tversky, 2004). Representational primitives are the properties highlighted by the construction parameters by means of which we identify a referent.

Depending on the identified construction rule, we isolate different representational primitives. They are relative and context-sensitive. Consider a picture of a red square. If one takes it as exemplifying red, the shape is nonrepresentational. The colour is irrelevant if one takes it as exemplifying the square construction.

Being relative and context-sensitive does not imply being arbitrary. The meaning of indexicals is relative and context-sensitive, yet it is not arbitrary. With this in mind, we can hold that images are canonically decomposable.

If images are canonically decomposable, then there is no reason to deny that they are systematic. They clearly are. If I have an image of two objects o_1 and o_2 where o_1 is left of o_2, I can easily imagine that o_1 is right of o_2. If I have an auditory image, I can imagine the same music track with different paces and pitches. Understanding one triangle diagram implies understanding any other. Moreover, systematicity is a key concept in understanding how images work. Let us recall the concept of natural generativity. One of the main tasks of any theory of depiction is to explain how understanding one member of an iconic system involves understanding any other member of the system. The concept of natural generativity mirrors the concept of systematicity.

However, the systematicity of images cannot be explained in the same way as the systematicity of propositional representations. Why is that so?

According to the classical explanation (e.g. Fodor, 1987; Fodor and Pylyshyn, 1988), propositional representations are systematic, for they have a logical form. Propositional representations are decomposable into discrete atomic formulas that contribute to the truth-conditions of the whole representation. For instance, the meaning of the sentence 'John is bald' can be derived from the meaning of its constituents. The idea is that every part is an atomic unit of a truth-evaluable representation.

Moreover, propositional representations have combinatorial syntax, which determines the place of their parts in this structure and the logical combinations of content constituents. For instance, the sentence 'John loves Mary' consists of the arguments 'John' and 'Mary', and the relation R of loving. Exhibiting the thought that John loves Mary involves understanding R, which, in turn, makes it possible to exhibit the thought that Mary loves John.

In contrast, images are deprived of logical form. They cannot be decomposed into atomic units of truth-evaluable representations and lack combinatorial syntax. Thus, we need an alternative explanation of the systematicity of images.

In the 2-D model, understanding iconic content involves knowing the construction rules by means of which we identify a referent. Recognition-based identification has been defined in terms of localizing the construction invariants connected with knowing the construction rules that determine permissible transformations of the constructed object. The idea is that we can think of the systematicity of images in terms of preserving construction invariants. Two representations are systematically co-related with each other if they preserve the same construction invariants. Let me illustrate this with two cases.

Think about a triangle diagram. If I understand how to construct one kind of a triangle diagram, I understand how to construct any other kind since I understand a rule which determines what properties have to be preserved in those constructions. Moreover, suppose I hold that I know how to construct a right triangle diagram, but I do not recognize the construction of an obtuse triangle diagram. In that case, it indicates that I do not understand triangle construction.

Let us consider an image of two objects o_1 and o_2 being left and right, respectively. Understanding that o_1 is left of o_2 implies that I understand that o_1

can be right of o_2, since they preserve their locations when the reference point is switched. Additionally, if I hold that I understand the image where o_1 and o_2 are left and right, respectively, but I cannot recognize the content of an image with the switched reference point, then I do not understand the first image.

Moreover, by manipulating construction parameters, one can indicate preserved properties. For instance, in the mental rotation tasks, we can spatially transform a figure, then check whether its shape fits another. By the same token, we can change a music track's pace and pitch, preserving the sound sequence. The shape of the spatial figure and the sequence of sounds are the invariants of these constructions.

The manipulation of the construction parameters can be cognitively productive. Think about a freehand drawing. It is based on searching for the best measure to identify a referent. By determining the parameters of the drawing, we identify different construction invariants, which, in turn, localize different properties of the desired object.

What is more, the result of such manipulation cannot be deduced. According to the measurement-theoretic account, images like rulers and balances identify the results of the measurement procedures. By systematically manipulating construction parameters, we test the consequences of some background assumptions involved in image production and interpretation. These consequences cannot be inferred. If they could be, any measurement would be reducible to inferential procedures.

Thinking with images is, using Dennett's metaphor (2013), 'turning the knobs'. It is about localizing objects, properties and events within some measurement set-up. It involves finding out what would change if the construction parameters were different. For instance, what would change if we spatially rotated an object in the mental rotation tasks or how would the world look if we reversed its colours.

To sum up, images are systematic and decomposable. They are decomposable into parts highlighted by construction parameters. They are systematic by means of preserving construction invariants. We do not need to introduce propositional structures to explain the systematicity and compositionality of images. Images are systematic and decomposable in their own right.

Application 1: Representational format

In the previous sections, I explained how the 2-D model addresses the epistemological, semantical and metaphysical challenges. In this and the next section, I describe the metaphysical landscape that can be seen from the measurement-theoretic account of images. I discuss two problems: the problem of the mental representation format and the nature of mental imagery.

A mental representation format is the way information is organized in mind. A discussion of mental representation formats addresses how the information in our mind is stored and processed. It concerns the structure of representations interpreted as a set of representational primitives and combinatorial principles. Thus, to describe a representational format, one has to describe the structure of representation: the primitive elements and the possible operations that can be carried out with them.

Following the influential tradition, we distinguish between analog and digital formats of representation. According to Goodman (1976), to be an analog representational system means to be a dense representational system. An example of a dense representational system is an old-fashioned clock that represents time continuously, unlike a digital clock that represents time discretely. Moreover, an analog representational system is relatively replete. A representational system is relatively replete if, in comparison with other systems, many features of its members are relevant to determining what they represent. The system of old-fashioned analog clocks is not replete since only the position of the clock's hands matters. In comparison, in the case of images, such features as colour, shape and size are relevant. However, for reasons that will not be covered here (e.g. Kulvicki, 2006; Maley, 2011), it is doubtful whether Goodman succeeded in adequately explaining analog and digital formats of representation.

In the last fifty years, the distinction has been variously interpreted and explicated (e.g. Fodor and Pylyshyn, 1981; Haugeland, 1998; Lewis, 1971; McGinn, 1989; Peacocke, 2019). Across those approaches, digital representations are generally understood to be discrete entities. Numerals provide a good example. 0 and 1 are discrete because they indicate distinct and separable entities. For every representational token, it is clear which type it instantiates. In contrast, analog representations do not admit definite type-identity. For example, the colour value of a given colour patch is measured on a continuous rather than a discrete scale (Dretske, 1981). Iconic representations are believed to be analog structures, which means that they are indiscrete. The structure of propositional representations is believed to be based on discrete structures.

An alternative way to interpret the analog–digital distinction can be framed in terms of constraints that the representational system puts on representational content. Different representational systems put constraints on the content a representational system can carry and the range of possible transitions between different contents. So, for example, an analog representation can represent a magnitude value but not an integer (Beck, 2015) and so on.

If one wants to explain why some representational systems put constraints on representational content, one has to describe the features of the representational structure. For instance, one can explain why the Arabic numeral system is preferred over the Roman numeral system by pointing out the fact that the Roman numeral system does not include a zero. Analogously, iconic and propositional representational systems put constraints on their representational content. A theory of representational format should explain where these constraints come from.

More generally, the question is what the concept of the mental representation format should explain and how it can do that. Here, I claim that the problem with the analog–digital distinction is not that it is not clear. Even if it were clear, it would still be doubtful whether it could explain what it should explain, namely, the difference between iconic and propositional representations.

There are at least two ways of describing the difference between propositional and iconic representations. The first way refers to the idea of the informative richness of images (e.g. Kitcher and Varzi, 2000). According to this criterion, iconic representations display more information than propositional representations, making images more effective in some reasonings (e.g. Larkin and Simon, 1987). The analog–digital distinction seems to explain this difference, for the analog representations consist of fuzzy sets of information, making iconic content vague and rich. In contrast, digital representations consist of classical sets of information, which makes propositional content more precise but informatively poorer.

However, a more profound understanding of the difference between iconic and propositional content can be described in terms of their relation to truth. Images are not truth-evaluable and lack logical form. In contrast, propositions can be true or false and express logical relations. A full-fledged theory of the mental representation format should explain what properties of the representational structure are responsible for these constraints put on the content.

Does the analog–digital distinction help us understand the difference between iconic and propositional representations characterized in terms of their relation to truth? It seems that it does not, as the iconic-propositional distinction does not overlap with the analog–digital one (e.g. Peacocke, 2019).

If we characterize content in terms of its relation to truth, then the distinction between analog and digital format is irrelevant for determining whether we are dealing with a representational system that is iconic or propositional. As von Neumann (1958) demonstrates, propositional operations, such as computations, can be functions of digital and analog processes. Propositional representations can be encoded in either digital or analog format. And although it does not show that the distinction between analog and digital representational systems is useless, it demonstrates that this distinction is insufficient for distinguishing images and propositions.

The measurement-theoretic account offers an alternative description of the representational format. In the 2-D model, images are measurement devices used to localize the depicted object. This feature generates some consequences for our thinking about the representational format. In the measurement-theoretic perspective, at least two marks distinguish the structure of iconic and propositional representations. The first concerns the problem of context sensitivity. The second is the problem of predication.

First, mechanisms of information processing in iconic representations are domain-specific. They are domain-specific if the operations defined for the structure elements depend on the application area. If the operations do not depend on the area they are applied to, the relevant mechanisms are domain-general. Such is the case of the rules of addition; regardless of what one is adding, the rules are the same. In contrast, measurement devices are domain-specific as measurement methods depend on what and how you measure.

Similarly, processing iconic representations depends on the modality and vehicle of representation.[4] For instance, visual and gustatory representations of wine belong to two different systems of representation. Moreover, there is a significant difference between imagining tasting wine and actually tasting it. The information is processed differently, depending on what and how it is represented.

In contrast, propositional representations are domain-general. The way information is processed is insensitive to the modality and vehicle of representation. For instance, the proposition WINE IS RED remains the same regardless of whether it is expressed in speech or writing, English or German.

Second, unlike propositions, images are not predicative. When I hold that snow is white, I attribute the property white to the object snow, where the terms 'snow' and 'white' work as arguments of a predicative function expressed in the proposition.

The predicative nature of representation distinguishes the proposition from an image. The proposition carries the denotations of the terms into a truth-value (Rescorla, 2009b), while the image of the white snow does not. What properties of the representational format can explain this?

For one thing, predication is based on a compositional and combinatorial mechanism. The compositional and combinatorial character of a propositional system means that if I have the propositions SNOW IS WHITE and A TRIANGLE IS RED, I can form the structurally similar propositions SNOW IS RED and A TRIANGLE IS WHITE. Moreover, if I know that the proposition SNOW IS WHITE is true, I can infer that the proposition SNOW HAS A COLOUR is also true. This means that an output of one operation of predication can be an input of another higher-order operation. These operations are hierarchically organized (Camp, 2018). From the proposition SNOW IS WHITE, I can infer that snow has a colour, but from the proposition SNOW HAS A COLOUR I cannot infer that snow is white.

For another, the output and input information comes in discrete chunks. Consequently, the proposition SNOW IS WHITE attributes the property WHITENESS and no other property to snow. It says nothing about the hue of the colour or the shape of snow, for there is a one-to-one correspondence between a vehicle and content. Every chunk of information needs a separate vehicle to be expressed. Thus, propositional representations are hierarchically organized and based on a structure made of discrete chunks of information.

In contrast, iconic representation is neither hierarchically organized nor based on discrete chunks of information. Hierarchical organization of representational structure means, for example, that the proposition SNOW IS WHITE implies SNOW HAS A COLOUR but not the other way round. In the case of iconic representation, both pieces of information are processed simultaneously. An image of white snow represents both that snow is white and that snow has a colour.

The non-hierarchical organization of information is often confused with the holistic nature of the components of the iconic structure. These two concepts, however, have to be separated since we can have non-hierarchical information architecture based on non-holistic components. For instance, parallel computing is non-hierarchically organized and based on discrete chunks of information.

The holistic nature of representation indicates the relation between the content and the vehicle of representation. To my knowledge, the first person to draw attention to this idea was Twardowski (1965), who ascribed the feature of concreteness to imaginings. He understood concreteness as the combination of multiple properties in a single representation. Similarly, the concept of holism is understood (e.g. Green and Quilty-Dunn, 2021; Kulvicki, 2020) as the thesis that multiple pieces of information expressing the content of a representation are assigned to the same vehicle of representation.

This means that there is no separate vehicle for every chunk of information corresponding to different representational properties. In other words, there is no one-to-one correspondence between parts of the information and parts of the vehicle. For instance, the part of an icon that represents the colour of a triangle is the same part that represents its shape.

In the 2-D model, the non-hierarchical information architecture and holistic nature of representational primitives result from how images represent the world. Unlike propositions, images represent by exemplifying the rules of construction which identify the represented object. In short, images are a kind of sample of the represented object.

Being a sample suggests that images have to be able to exemplify many properties at once. Depending on the construction rule we pick out, different representational properties are highlighted. Yet, there is no principled way to determine the property that is represented by the image. That requires that the vehicle carry more than one piece of information at once.

If there is no one-to-one correspondence between the vehicle and information, then information processing cannot be hierarchically organized. For instance, a triangle diagram represents the triangle as being right and red, simultaneously. Such information is not inferred. It is displayed by the same vehicle of information.

Thus, the format of iconic representations is domain-specific; it processes information in a non-hierarchical manner and is based on holistic components. The structure of propositional representations is domain-general; it processes information hierarchically and is based on discrete chunks of information.

Does this quality of representational format allow us to explain the constraints put on the content of images and propositions? It does. First, displaying multiple pieces of information by one vehicle of representation explains why we hold that images are informatively rich. Iconic content is informatively rich, for it is able to express more information than propositions, which are constrained by the one-to-one correspondence between the vehicle and the information.

Second, being truth-evaluable and having a logical form requires propositional structure. It involves a special kind of relation between the vehicle of representation and the content. To speak of truth, we need to be able to distinguish atomic formulas that can satisfy T-schema. That requires a one-to-one correspondence between the vehicle and the information. Similarly, to infer from a is F and if a is F, then a is G that a is G, one has to be able to assign distinct content to the vehicles of representation expressed by logical variables. Moreover, the chunks of information have to be hierarchically organized. From the thought that a is G and that if a is F, then a is G I cannot infer that a is F. Propositional representations are hierarchically organized.

In contrast, images are organized non-hierarchically. The information is processed simultaneously. When I see a picture of white snow, I see that it has a colour; when I see a colourful picture, I can see the specific colour of the picture.

Moreover, propositions require a representational structure that can abstract away from the nature of the vehicle of representation. For instance, the reasoning if *A* then *B* and *A* then *B* is correct regardless of whether it is conducted mentally or on paper. Propositional representations meet this requirement since they are domain-general. In contrast, iconic representations are domain-specific. The nature of the iconic representation vehicle affects how information is processed. For instance, tasting wine and imagining tasting wine are informationally two different representations.

To sum up, iconic representations are not truth-evaluable and lack logical form. These facts are easier to understand if we hold that iconic representations are domain-specific, that they process information in a non-hierarchical fashion and that their structure is based on holistic components. In contrast, propositional representations are domain-general, they process information hierarchically and their structure is based on discrete elements.

Application 2: Mental imagery

Mental imagery is usually understood as a faculty of bringing about mental images and seeing with the mind's eye (e.g. Kosslyn et al., 2006; MacKisack et al., 2016).[5] However, asking about the nature of mental imagery is tricky.

In cognitive psychology, the nature of mental imagery has been the subject of the so-called imagery debate. However, there has never been one understanding of the debate's subject. Instead, there have been at least three parallel debates. What started as a discussion on epistemological functions of mental images (Paivio, 1963; Shepard and Metzler, 1971) soon turned – mainly thanks to the computational character of research in early cognitive science – into a discussion on the format of mental representation. On the one hand, pictorialists (Kosslyn, 1980) have held that the format of mental images is perceptual-like. On the other, descriptionalists (Pylyshyn, 1973) have argued that mental images are formed out of structured descriptions. Most descriptionalists have argued that mental images are epiphenomena of some internal language-like processes. From the beginning of the 1990s, the discussion has focused mostly on neurobiological mechanisms responsible for generating mental images. None of these debates has direct implications for others.

Nowadays, the imagery debate is mostly considered dead (Pearson and Kosslyn, 2015). In cognitive psychology, there is a consensus that descriptionalists fail to explain the perceptual nature of mental images. Moreover, it is held that mental imagery is a form of weak and off-line perception (e.g. Pearson et al., 2015). According to the weak perception theory of mental imagery, mental images are a form of less vivid perception not triggered by sensory input.

The claim that descriptionalism is a dead end seems to be noncontroversial. Moreover, there is no doubt that imagining is a form of perceptual experience and that the neural correlates of perception and mental imagery are common to a significant degree (e.g. Kosslyn et al., 2006; Nanay, 2015; Pearson et al., 2015).[6] There are, however, doubts regarding the consequences of these claims. Primarily, the weak perception theory of mental imagery is philosophically dubious.

First, although mental imagery and perception share the mechanisms responsible for generating representations, there is an asymmetry between the content of mental imagery and perception. It is believed that the content of mental images is displayed as already interpreted (e.g. Chambers and Reisberg, 1985; Hinton, 1979; Ishai et al., 2000; Ittelson, 1996; McGinn, 2006; Reisberg, 1996; Reisberg and Heuer, 2005; Sartre, 1962; Scaife and Rogers, 1996; Slezak, 1995). In contrast, we can always reinterpret the content of perception.

The most striking illustration of this fact is the problem with reinterpreting the so-called bistable figures (such as the duck-rabbit picture) in mental imagery. Although reinterpretation of mental images is possible (e.g. Reisberg and Chambers, 1991; Finke et al., 1989; Peterson et al., 1992), it is more cognitively loaded and significantly less efficient than in perception (Mast and Kosslyn, 2002; Kamermans et al., 2019). For instance, in the studies of Reisberg and Chambers (1991), subjects were told to remember and imagine a figure with the same geometry as the state of Texas rotated 90° clockwise. Even when they imagined rotating the figure mentally so that it assumed its canonical orientation, they were still unable to discover that it represented Texas. In contrast, the task was trivial when subjects were told to rotate the image being seen.

The asymmetry between imagery and perception when interpreting bistable figures cannot be explained by holding that mental images are displayed in the mind's eye too briefly, which might suggest that imagery is affected by memory limitations (see Kosslyn, 1994; Kosslyn et al., 2006). For one thing, reinterpreting the perceptual content does not have to be time-consuming. Often, we can switch between two interpretations instantaneously. For another, the time that affects the efficiency of reinterpretation tasks is the same time that suffices to solve mental rotation and mental zooming tasks in Kosslyn's classic research on mental imagery (e.g. Kosslyn et al., 1978). Therefore, the time for which information is available does not seem to be a relevant factor.

Second, suppose that we respond to the first doubt by holding that mental imagery is less vivid than perception. However, that is a position that we do not want to keep if we want to defend the imagistic theory of thinking. For one thing, the weak perception theory of mental imagery appears to blur the distinction between thoughts and perception and is vulnerable to arguments from cognitive impenetrability of perception (Cavedon-Taylor, 2021; Firestone and Scholl, 2016; Raftopoulos, 2009). For another, it seems to take us back to the neo-Lockean positions we previously rejected (see Slezak, 2002a). In fact, the weak perception theory of mental imagery is no more philosophically attractive than the Lockean theory of ideas, which has been a philosophical dead end for over three hundred years. If we hold that mental imagery entails weak perception, then either exercising mental imagery is not a thinking operation or speaking of imagistic theory of thought is senseless. Both alternatives are unattractive.

Where do we stand? The situation resembles the one described by Goodman and Elgin (1988) as the dilemma of mental imagery. Either we accept the weak perception theory of mental imagery and leave out the questions asked by philosophers and other troublemakers, or we just dismiss the talk of mental images, holding that they are just the epiphenomena of some more substantive cognitive mechanism. I think that both horns of the dilemma should be rejected.

The problem we face stems from overusing the photographic metaphor in thinking about mental imagery. According to this metaphor, mental imagery is like a camera, and mental images are like photographic snapshots displayed before the mind's eye. The force of the photographic metaphor results from our attachment to the Traditional View. If images are copies of the world, then mental images are copies of perceptual experiences. That is, however, where we get stuck in conceptual confusion. There is no Cartesian cinema in the head; some inner eye does not see mental images.

According to Block (1983b), the photographic model of mental imagery should be replaced by the one in which mental images are based on mechanisms involving constructive processes. Mental images are more like drawings rather than photographs. However, Block has never been specific enough to explain what such processes can be.

In the measurement-theoretic account, the drawing model of mental imagery can be explicated as follows. According to the 2-D model, images are measurement devices. They exemplify the construction rules by means of which we identify the depicted objects and events. Iconic content localizes objects and events in some physical or logical space by determining the parameters of the space. Knowing the iconic content is knowing the construction rules that set these parameters.

The general idea is that we can distinguish between two ways of understanding the concept of the mental image. One refers to having a conscious, perceptual-like experience of an object.[7] In the 2-D model, it is a vehicle of representation that exemplifies iconic content. Let us call this the experiential account of mental images. The other refers to the iconic content that is exemplified by this experience. Let us call this the constructive account of mental images. Having a mental image in the constructive sense is knowing the iconic content. According to the measurement-theoretic account of images, this means knowing the construction rules through which we identify the object. Just like having a concept does not imply producing it, having a mental image in the constructive sense does not imply producing the experience of mental images.

Constructive understanding of the nature of mental images has at least three metaphysical consequences for our understanding of what mental imagery is. First, mental imagery as a faculty of bringing about mental images is not, as it were, a moving theatre where mental images are displayed before the mind's eye. The cinema-in-the-head idea of mental imagery results from taking the experiential understanding of mental images as constitutive of what mental images are. Mental images, however, do not have to be experienced. Mental images as conscious experiences are exemplifications of a more basic iconic mechanism.

According to the constructive understanding of mental images, having a mental image is an ability to perform certain activities (Goodman, 1984; Goodman and Elgin, 1988). In the 2-D model, to have a mental image is to know a construction rule that is a measure by means of which we identify objects and events. For knowing this measure, however, it is irrelevant whether you have an object that exemplifies it. It is crucial to have the skill of measuring. For instance, you do not need a ruler to measure distance. If you know how to measure distance, you can use anything that can help you, for example, a stick.

Mental imagery is the skill of applying construction rules to identify objects and events in some space. It involves determining the parameters of some space to arrive at the constructed object and recognize it. Mental images as conscious experiences are exemplifications of these skills.

Does this mean mental images as conscious experiences are irrelevant to mental imagery? It depends on how you understand this question. Thinking of mental imagery in terms of the skills of producing mental images does not imply that it actually creates them. By the same token, when thinking in words, you do not have to produce these words. For instance, you can have a thought and search for the words to express it. What is crucial is the skill to identify the linguistic representations that correctly express this thought.

Mental images as conscious experiences are products of mental imagery taken as a procedural knowledge of construction rules. This knowledge manifests itself in the operations of constructing and interpreting the mental image. However, it can also be manifested by any intelligent use of pictures, such as skilful use of diagrams.

Taking (mental) images as products of procedural knowledge does not imply that they are irrelevant for exercising such a skill. If one cannot identify the construction rules represented by an image, regardless of whether the image is mental or extramental, then one does not have this procedural knowledge. In this sense (mental) images are necessary for mental imagery.

To illustrate this, let us once again consider mental rotation tasks. The idea is that subjects manipulate the parameters of perceptual space by rotating the figure in order to localize construction invariants. Finding these invariants solves the task of identifying two similar objects. The task cannot be solved without mental imagery since it involves procedural knowledge of how to determine these parameters, for example, setting the direction of rotation.

However, having such procedural knowledge is dissociated from having an actual experience that exemplifies it. The operation of rotating a figure can be visualized in the form of a conscious experience, but it does not have to be. We do not have to represent the skill as a conscious mental image to exercise it. We have to be able, however, to recognize the identity of figures when we visualize them or display them in pictures.

Second, if mental imagery does not have to be conscious, then we can speak meaningfully about unconscious mental images. Let us recall the phenomenon of aphantasia, which is the lack of the ability to form mental images (e.g. Keogh and Pearson, 2018; Milton et al., 2020; Zeman et al., 2015, 2020). This phenomenon appears to be theoretically problematic only to the experiential accounts of mental imagery. From the measurement-theoretic perspective, the phenomenon of aphantasia does not seem to be particularly mysterious. What is lacking is not mental imagery but the awareness of having mental images.

Being unaware of a mental image does not imply that one lacks the skills to apply construction rules. For example, aphantastics can recognize identical figures represented in pictures which indicates that they do not lack mental imagery.

Third, the relationship between perception and mental imagery seems much more profound than the weak perception theory suggests. Mental imagery is not some less vivid perception. It is a skill of perceiving based on procedural knowledge of construction rules. It involves the skill of identifying the parameters of perceptual space in order to localize an object or event.

Thinking of mental imagery in terms of perceptual skills allows us to explicate the relation between perception and mental imagery. For one thing, it allows us to acknowledge the perceptual nature of mental images. If mental imagery is a skilful use of perception, then the neural mechanisms of perception and mental imagery should be shared, which is indeed the case. Moreover, if the format of the iconic representation is domain-specific, then we should not expect to have one mechanism responsible for mental imagery, which is also the case. There is no common mechanism of mental imagery for all modalities.

For another, we can explain the asymmetry in interpreting bistable figures in imagery and perception. If mental images are representations of measures, then we expect their meanings to be fixed. They are not open to interpretation, in the same way as the meaning of a standard metre is not open to interpretation. Images set the conditions of interpretation by exemplifying the construction rules by means of which we identify objects. In contrast, perceptions are open to interpretation, for it is always possible to ask which measure should be applied to interpret them. Depending on different measures, we highlight different perceptual properties.

Moreover, interpreting mental imagery in terms of a set of skills gives us a better picture of the relation between perception and thinking than the neo-Lockean approaches. On the one hand, it offers a hint on how to think about the role of mental imagery in perception. A good illustration of such a role is the so-called amodal completion. It is the common phenomenon of perceiving the parts of perceived objects from which we have no sensory stimulation (e.g. Nanay, 2018).[8] For example, when seeing a cat behind a fence, we amodally complete the parts of the cat that are occluded. Importantly, amodal completion is a perceptual phenomenon. It happens very early in the sensory cortices (Thielen et al., 2019). In a sense, we see the occluded parts that are not triggered by the retinal input.

Now, if we think about mental imagery in terms of perceptual skills, then amodal completion is a form of mental imagery. The procedural knowledge of how to construct an object amodally completes this object in perception. For instance, seeing a part of a cat hidden behind a fence involves applying the construction rule by which we reconstruct the whole object in mind. By the same token, when seeing an incomplete or badly drawn triangle, we complete the figure, for we know how to construct triangles.

On the other hand, the measurement-theoretic account appears to be invulnerable to objections from the thought-perception border. According to this objection, there is no room for any representation to mediate between thoughts and perception since they possess mutually incompatible properties. However, mental imagery is not a representation that mediates between thought and perception. It is a perceptual skill of applying construction rules that results in recognition-based identification of objects.

Recognition-based identification triggered by knowing specific construction rules is a phenomenon that is both perceptual and cognitive (Abid, 2021). It is perceptual since it involves perceiving the recognized objects. It is cognitive since it involves a cognitive operation of applying the rules of construction. And yet, it is not a belief-based operation. It does not involve having any beliefs about the recognized objects. Having beliefs involves possessing concepts and the language to express them. Perceptual recognition is a skill that precedes having beliefs.

Let us go back to amodal completion. It is both perceptual and cognitive. From the neuroscientific point of view, we see unperceived objects. However, it is not the operation that has to involve any beliefs. I can amodally complete a perceptual object I have no beliefs about. Instead, it is a matter of a thoughtful exercise of skill of recognition.

To sum up, by distinguishing between the experiential and constructive understanding of mental images and interpreting mental imagery in terms of constructive skills, we can solve the dilemma of mental imagery. On the one hand, mental imagery as a Cartesian theatre displayed before the mind's eye is only another instantiation of the myth of Ryle's (1949) ghost in the machine. On the other, mental imagery is a skill of applying construction rules to identify objects in some space. In this second understanding of the term, mental imagery is an empirically well-grounded phenomenon that can be studied by mental rotation, mental zooming tasks and so on. From the philosophical point of view, mental imagery seems to be no more problematic and can be studied with empirical methods. If some philosophical questions can be answered with empirical methods, the philosophical job is done.

Summary

In this chapter, I explain how the measurement-theoretic account of images gives us an insight into the nature of imagistic knowledge. Images exemplify measures that make it possible to understand the objects of our beliefs and theories by means of localizing these objects in some logical or physical space. I hold that iconic content does not require criteria of interpretation. In contrast, it sets these criteria. Images, like rulers and balances, exemplify the measures by which we recognize objects, properties and events. Finally, I demonstrate that images are systematic and decomposable in a different way than propositional representations. They are systematic and decomposable as they exemplify the rules of construction which identify construction parameters and invariants.

I show the measurement-theoretic account of images allows us to acknowledge the irreducible role of images in thinking. Images serve the function of localizing objects of our beliefs and theories. At the same time, imagistic and propositional representations are bound together. Although images can localize objects, enabling their recognition-based identification, they cannot describe them. We need propositional structures to make images meaningful, that is, to place them within our beliefs and theories.

Last but not least, I demonstrate the consequences of the measurement-theoretic account of images for our thinking about the mind. First, the account allows us to rethink the concept of the representational format. Iconic representations are domain-specific, they process information in a non-hierarchical fashion and their structure is based on holistic components. Second, it offers us a fresh perspective on the idea of mental imagery. According to the measurement-theoretic account, mental imagery is a measuring skill that determines the parameters of perception.

8

Conclusion

In this book, I have been trying to solve the philosophical problem of determining the nature of thinking with images. The problem is how to reconcile two separately legit but mutually incompatible theses. On the one hand, we are justified in holding that images are irreplaceable in cognitive practice. On the other, the way we think about thinking makes no room for imagistic thoughts. Images seem to be instrumental and peripheral in thinking.

The problem stems from two sources. First, it is based on our assumptions regarding thinking. We usually hold that thinking is propositional and action-based. Thinking is an operation carried out on propositional structures that are expressed in language-like representations. It builds the so-called Received View on the nature of thinking.

Second, when thinking about images, we usually adopt a version of the so-called Traditional View, according to which images represent objects by means of resembling them. Adopting the Traditional View, however, makes images vulnerable to Wittgenstein's argument from the indeterminacy of content and Frege-Davidson's argument from lack of logical form. Thus, in contrast to propositional representations, images seem unable to deliver the theory of knowledge and the content of thought. They cannot explain the systematic and decomposable structure of thought. Therefore, imagistic thinking seems to be a contradiction in terms.

And yet, we cannot simply reject the Received and Traditional Views. They both express intuitions that are central to our thinking about the nature of thoughts and images.

Throughout the book, I have been trying to address these problems by taking the so-called operational approach, according to which the question 'What is thinking with images?' can be taken as the question of possible operations carried out with the help of images. Explaining the nature of imagistic thinking means pointing at the conditions that must be met if such operations are possible. Thus, I investigate what images do and try to find something that does it.

To find out what images do, I analyse the functions of images. I am doing so by presenting two case studies. The first one concerns the knot diagrams in mathematics. With this example, I study the idea of the content of diagrams. To explain the way diagrams represent, I introduce the concept of construction. It refers to the procedures of arriving at a goal by determining the parameters of some logical or physical space. The general idea is that diagrams represent objects and events by representing the ways these objects and scenarios can be constructed.

By introducing the concept of construction, I am able to offer an alternative account of iconic content. In contrast to the Traditional View, I hold that resemblance works as an explanandum but not as an explanans. I hold that the concept of construction can explain why images resemble their objects, avoiding the objections directed at resemblance-based semantics.

In the second case study, I analyse the black hole picture. To explain the way this picture represents its object, I introduce the concept of recognition-based identification. I argue that recognition-based identification is different from demonstrative reference and descriptions. Recognition is based on identifying the construction invariants.

Based on the concepts of construction and recognition-based identification, I introduce the so-called two-dimensional model of iconic reference. According to this model, images denote their targets and exemplify the rules of construction by means of which they reidentify the referent.

I argue that the best way to explain these properties of iconic reference is to think of images in terms of measurement devices. Just like rulers and balances, images represent the ways of localizing objects and events by determining the parameters of some space. In contrast to the Traditional View, images are not copies of the world. They are the measurement devices we use to localize the properties of the world.

Next, I demonstrate that the measurement-theoretic account of images offers a theory of imagistic knowledge and content. In the 2-D model, imagistic thoughts are systematic and decomposable. Thus, thinking of images in terms of measurement devices meets the requirements of a theory of thought. Moreover, it is a kind of thinking that is irreducible to propositional thinking.

Last but not least, I show the metaphysical consequences of the measurement-theoretic account of images for our thinking about the format of mental representations and mental imagery. I hold that we can replace the analog–

digital distinction with one where the structure of iconic representations is domain-specific, non-hierarchical and based on holistic components. In contrast, propositional representations are domain-general, hierarchical and based on discrete elements.

Finally, according to the measurement-theoretic account, we can abandon the Cartesian view on mental imagery. In the 2-D model, mental imagery is a skill of applying construction rules that determine the parameters of perception.

To sum up, I have been arguing that the imagistic theory of thought can be made meaningful if we think of images in terms of measurement devices. In the measurement-theoretic perspective, images can be seen as irreducible to propositional representations. This does not imply, however, that images are more basic than propositional representations or that they can ground the meaning of propositions. Iconic and propositional representations have distinct functions, yet they are entangled in thinking. Thus, if one asks whether we think with words or images, the answer would be that we think with both. Thinking is both iconic and propositional.

Does this imply that we have words or images in the mind? No, it does not. Thinking is an operation that can be expressed with either words or images. Words and images are necessary for thinking only in the sense that they are the only medium that can express these operations. Images and words are necessary for having thoughts, just as numerals are necessary for numbers. They represent some operations. In the case of images, these are the operations of construction and recognition. Propositions are indispensable for the operations involving truth values, such as inference or having beliefs.

One of the consequences of the ideas presented here is that knowledge is more than propositional knowledge and thinking is much more than the Fregean game in truth. In contrast to Frege, I hold that thinking involves much more than truth-evaluable operations. What this 'something more' is should be a philosophical topic, just as truth is. Even if this book does not succeed in presenting a coherent theory of non-propositional thinking, I hope that it presents the need for such a theory. I am convinced that the philosophy of images can play the same role in our understanding of the mind as the philosophy of language has played.

Although the book covers many topics, it does not cover all of them. I barely touch on the problem of scientific models and the arts. I deliberately omit the problems of metaphors and imagination. These issues require separate studies. I believe, however, that they can be interpreted in line with the 2-D model and incorporated into the imagistic theory of thought. Yet, any attempt to discuss these problems in one book would far exceed the reader's patience.

In the book, I have argued that although the idea that we think with images is intuitive, it has been rarely thought through. However, this does not mean that it has never been considered. In general, no idea presented here cannot be found in Kant, Peirce, Wittgenstein, Evans or Goodman. All I am presenting here is an attempt to think through the consequences of their ideas. All I am hoping for is that we can see some ideas clearer. That is the only thing a philosopher can fairly offer.

Notes

Introduction

1 I use the expressions 'thinking with images' and 'thinking in images' interchangeably.
2 See, e.g., Irvine (2014), Lakoff (1994), Lakoff and Johnson (1980), Langacker (1987). For a contemporary defence of an imagery-based theory of semantics, see, e.g., Ellis (1995), Lowe (1996), Thomas (1997) and Nyíri (2001).
3 For a philosophical and empirical defence of the existence of unconscious mental images, see, e.g., Church (2008), Nanay (2010, 2015, 2017, 2021) and Zeman et al. (2010, 2015). The discussion mostly concerns the phenomenon of aphantasia, which is a condition where one does not possess the ability to visualize images (e.g. Keogh and Pearson, 2018; Milton et al., 2020; Zeman et al., 2015, 2020) and form non-visual mental images (e.g. Dawes et al., 2020).
4 Mental and extramental images undoubtedly differ in many respects. The differences, however, should not obscure the similarities (e.g. Abell and Currie, 1999). Reasoning with the help of mental images is often performed interactively with external representations (e.g. Kirsh and Maglio, 1994). Forming an extramental image is not a post-hoc expression of some reasoning done with the help of mental images but an integral part of this reasoning (Sheredos and Bechtel, 2020). Moreover, an ability to form an external image assumes an ability to form a mental image; for example, knowing how to construct a triangle on a piece of paper is knowing how to construct a triangle in the mind, and vice versa (Maynard, 2011).
5 Most generally, this relation to a proposition can be described in Fregean terms as an act of 'grasping' thoughts, which refers to every act of discovering or understanding the meaning of a thought and which is equivalent to knowing the truth-conditions of a proposition. Grasping a thought is distinguished from evaluating the truth value of a thought, for knowing the conditions according to which a thought is true is not equivalent to knowing whether it is true.

Chapter 1

1 In fact, it is almost impossible to find a single empirical article without any reference to a graph or a diagram. According to the studies of Zacks et al. (2002), the number of graphs in scientific journals doubled between 1984 and 1994. Nothing suggests that this trend has not continued.

2 The concept of diagrammatic reasoning is vague, since there is no general agreement on the definition of a diagram (e.g. Pietarinen, 2016; Shimojima, 2001; Stjernfelt, 2007). A diagram is usually interpreted as a spatial representation of abstract pieces of information that enable us to infer the features of the represented objects through inspection of the spatial characteristics of the representation (Swoyer, 1991). Diagrammatic reasoning may be described in terms of a reasoning process based on manipulation and inspection of diagrams.

Chapter 2

1 In cognitive psychology, the belief that thinking is propositional and action-based is best expressed by Holyoak and Morrison's (2012) definition of thinking as the systematic manipulation of cognitive representations determining current or possible states of the world. In accordance with the Received View, thinking is most often operationalized as an act of reasoning and problem-solving (Evans, 2017). On the history of research on thinking in cognitive psychology, see, e.g., Dominowski and Bourne (1994).
2 An example of the epistemological argument from the content indeterminacy of images is Descartes's well-known chiliagon-argument. According to Descartes, an image of a chiliagon is indistinguishable from an image of polygon with 999 sides, and yet we have clear and distinct ideas of a chiliagon and a 999-gon. Thus, thinking cannot be imagistic. However, this argument is clearly invalid, since it does not imply that our thoughts cannot be imagistic. It demonstrates only that our perceptual capacities are limited. If we have enough time to draw a picture of chiliagon and a 999-gon and count the angles, then it is possible to distinguish a chiliagon and a 999-gon (see Dennett, 1990). In the imagery debate, contemporary versions of Descartes's epistemological argument (sharing the flaws of the original argument) are Dennett's argument (1981a) from a striped tiger and Armstrong's argument (1968) from a speckled hen.
3 Kulvicki's argument concerns pictures, but it can be readily extended to all imagistic genera.
4 The psychological instantiation of the pictorial fallacy is the so-called lack of pictorial competence. It is an inability to discriminate between objects and depictions of these objects which characterizes young children up to two years old. It manifests itself in toddlers' attempts to manually grasp the depicted objects as if the real objects were presented (see DeLoache et al., 2003).
5 Gregory (2020) argues that images are non-propositional but predicative. The Semantical Challenge shows that the predicative function is suspect.
6 According to Evans (1982), systematicity is constrained by semantic conditions of appropriateness. For instance, thinking that JOHN FELL INTO THE LAKE need

not entail a capacity to think that THE LAKE FELL INTO JOHN. However, even if a well-formed string is semantically absurd, that does not mean that it cannot express a thought. For one thing, we can entertain absurd thoughts. For another, absurd but well-formed strings can serve as a basis of inferences in logic. See, e.g., Camp (2004).

7 That does not mean that they lack construction rules. They are obviously rule-governed. That is why if one understands how to interpret one Venn diagram, then one understands how to understand another (e.g. Tversky, 2004). A representational system can have construction rules without syntactic structure. A syntax requires discrete symbols and logical rules that transform strings of symbols into well-formed formulas of the system.

8 Propositional and associationist theories of thought, just as the Received View, do not represent actual theories or a set of theories but certain ideal types or research programmes. Although they share some methods and questions, they are sets of related yet independent theses. In this general sense, associationist theories of thought have to be distinguished from historical associationist theories of mind such as Mill's and Hume's.

9 There are also intermediate strategies, such as Rescorla's (2009).

10 This distinction does not cover all possible accounts of thinking. There are also atypical positions, such as Ryle's adverbial theory.

11 To the best of my knowledge, Kant and Peirce were the first to recognize perceptual and spatial metaphors as complementary in their theories of thinking.

Chapter 3

1 In twentieth-century philosophy, some philosophers have seen no choice but to directly include this intuition in their theories of mind. The most significant examples were Russell (1921) and H. H. Price (1953). Both argued that a full-blooded theory of mind should include imagistic thoughts and that images are more basic than words. In a sense, images have a superiority over words, for they ground the meaning of language. As a consequence, they, rather tacitly, take imagistic thought to be a primitive and unanalysable phenomenon (e.g. List, 1981). In other words, they assume that the content of images is self-explanatory. However, if Wittgenstein's argument from content indeterminacy is sound, then images cannot be self-interpretable atoms of meaning.

We encounter similar problems in Arnheim (1969, 1980) and Wollheim (1984). Both try to find a description of the phenomenology of thoughts that is in accord with the imagistic nature of representations. However, even if some thought processes may appear as if they were imagistic, that does not mean that their real nature is imagistic. It only means that we need an explanation for why they appear imagistic. Given that they demonstrate that we think with images, it does not bring us any closer to answering the question of what thinking with images is.

In cognitive science, we encounter the same problem. It is well established that there are individual differences in thinking and people can be characterized in terms of being 'concrete' versus 'abstract' thinkers (e.g. Paivio, 1963; Paivio and Harshman, 1983; Kozhevnikov et al., 2005). There is also something to the observation that visual aids can facilitate the process of thinking. For instance, it has been frequently indicated that the same sort of 'visual' style of thinking fosters the creativity of scientists (e.g. Holyoak and Thagard, 1996; Johnson-Laird, 1998; Otis, 2015; Rocke, 2010; Shepard, 1978). However, we must bear in mind that we lack a clear-cut definition of thinking styles. This concept is most often understood very broadly. It refers to many different domains, such as cognitive styles, learning styles, personality types, and so on, which are not integrated within a general theory of psychological functioning. There is strong evidence that there are no unified visual and verbal cognitive styles (e.g. Kozhevnikov et al., 2005, 2010). Moreover, it is crucial to remember that the psychology of individual differences does not study thinking mechanisms but asks about individual differences in using different mechanisms depending on preferences and needs. Thinking styles are studied based on questionnaires and surveys. There is no way to study them using neuroimaging methods. The fact that there are individual differences in solving cognitive tasks does not imply that we were born with some inherent thinking preferences. It can mean that using images is a good cognitive strategy for specific tasks, and some individuals are more skilled in employing one strategy or another. Thus, referring to thinking styles does not explain what thinking with images is. It is only another way of describing the explanandum. A proper explanation requires analysing the source of individual differences in cognitive tasks. We should find out what kind of functions images can play and what cognitive purposes they are better for. Foremost, what has to be explained is why imagistic thinking strategies are sometimes better than non-imagistic ones.

2 Historically speaking, there is controversy as to who in fact represents the imagistic theory of thought. Berkeley appears to believe that all ideas are images (for the opposite view, see, e.g., Kasem, 1989). Hume believes something similar, though there are some doubts (e.g. Yolton, 1996). Locke's view is far from clear. It can be argued that we do not have to refer to images to understand the nature of Locke's concept of idea and that we can instead adopt an adverbial interpretation. See, e.g., Ayers (1991), Chappell (1994), Lowe (1995), Soles (1999), White (1990) and Yolton (1996).

3 In the light of grounding cognition in perception, arguing away the innate elements of the mind seems to be relatively less important. In the case of Locke, the point was that empiricism would be more convincing if there were no plausible

nativist account of knowledge that could challenge his position. In general, nativism is not incompatible with the imagistic theory of thought. For instance, in cognitive psychology, conjoining nativism with imagistic theory is instantiated by the 'core cognition' approach. It is based on the assumption that the human mind may be innately endowed with some basic, presumably iconic, cognitive systems that explain the capacity to learn exhibited by human infants (e.g. Carey, 2009, 2011; Spelke and Kinzler, 2007) which could help make sense of the apparently retrospective nature of the representations that underlie many looking-time experiments in the violation-of-expectancy paradigm (e.g. Baillargeon et al., 1985).

4 There are two groupings of evidence: behavioural and neurobiological. The behavioural evidence is based on experimental results indicating that tasks involving executing mental operations on perceptual or motor representations interfere with specific perceptual experiences or actions (e.g. Pecher et al., 2003, 2004). The neurobiological evidence shows that tasks involving mental representations operations recruit perceptual areas of the human brain (e.g. Beauchamp and Martin, 2007).

5 That does not mean that there are no cognitive effects on perception. For instance, perceptual interpretation of an ambiguous picture differs before and after information is given about what the picture is of (e.g. Fazekas and Nanay, 2017; Teufel and Nanay, 2017). It does not, however, influence the visual mechanism of picture-perception.

6 The most striking findings come from research on number sense (e.g. Cantlon, 2012; Dehaene, 2011; Izard et al., 2009; Machery, 2007). There is also a number of neurobiological findings that indicate the amodal nature of concepts (e.g. Bedny et al., 2008, 2012; Mahon, 2015).

7 The operational approach is not a kind of a functional explanatory strategy, since it does not determine any causal roles. The explanatory goal is to determine the necessary conditions of some logical operations.

8 This is a paraphrase of the strategy David Lewis took with respect to explaining the nature of meaning: 'In order to say what a meaning is, we may first ask what a meaning does, and then find something that does that' (1970, 22).

Chapter 4

1 I hold that knot diagrams are necessary to represent the ways of constructing knots. This does not mean that every property of a knot is representable by a knot diagram. For instance, knots have properties represented only by continued fractions and polynomials.

2 Notably, in the commentary to Euclid's Elements, Proclus (1992, 159) distinguishes the concept of construction from the concept of proof. The latter refers to a process of drawing inferences by means of rigorous reasoning from propositions to conclusions. Hilbert showed that Euclidean constructions are dispensable in proofs; however, they were never supposed to be a kind of proofs.

3 Tichy (2012) uses the concept of construction to explicate Frege's theory of sense and reference. It is, however, irrelevant for our purposes.

4 Intuitively, it may seem offensive to interpret algebraic formulas as instantiations of diagrams. Yet, we should not overestimate the power of this intuition. Historically speaking, the prevailing belief was rather the opposite. For instance, Kant understands '7 + 5' in terms of a kind of construction, not in terms of a power of the sets (e.g. Shabel, 2003). According to Kant, algebra 'achieves by a symbolic construction equally well what geometry does by an ostensive or geometrical construction' (Kant, 1998, A717/B745). The same intuition can be found in Peirce. He holds that (CP 2.282) 'every algebraical equation is an icon, in so far as it exhibits, by means of the algebraic signs (which are not themselves icons), the relation of the quantities concerned'. According to Peirce, basic units of algebraic constructions are not iconic. Yet, they do not have to be. What makes algebraic equations diagrams is that they demonstrate how we can construct magnitudes. Thus, for Peirce (CP 3.419), 'algebra is but a sort of diagram'. Let us follow this tradition and see where it can lead us.

5 The concept of construction is defined very broadly. In general, everything can be taken as an effect of a construction. Yet, although everything is a product of a construction, not everything is a representation of a construction. The term '7 + 5' can represent a procedure of arriving at '12', as well as be an effect of a construction operation representing the proposition 7 + 5 = 12.

6 It may seem natural to interpret the content of diagrams in terms of representing Platonic ideal objects. In this book, I do not offer a conclusive argument against Platonism, although I take this view untenable. However, interpreting the content of diagrams in terms of the rules of construction offers an alternative to Platonism. Consider the concept of the inflation growth. According to the view presented here, inflation growth is representable in diagrams. However, inflation growth is not an ideal object the diagram resembles. The inflation growth is an effect of construction procedures based on the application of the ordering function to the inflation rate and time. Does this mean that inflation growth is unreal? No! It means that inflation growth is not an (ideal) object like a table or chair. However, providing an argument against Platonism in favour of more credible ontology of mathematical objects would require a different book.

7 If we think of diagrams as representations that show how to construct relations within some space, we can abstract from the kind of space we are in, for the properties of the space do not determine the intrinsic properties of the diagram.

Thus, the space can be a physical or logical space. For instance, the inflation diagram localizes the values of time and inflation rate and determines the relation between them. The term '7 + 5 = 12' determines the relation between some magnitudes. Geometrical diagrams are only a subset of a more general phenomena.
8 The distinction between type-diagram and token-diagram is Peircean in origin (e.g. *NEM* 4.315).
9 Possessing such disposition does not imply that it will be actualized. For instance, one can have no talent for drawing diagrams. In the same way, if I speak in some language, I have a disposition to read in this language. However, such disposition cannot be actualized, if I do not know the alphabet.
10 Note that I claim that diagrams can help discover a proof strategy (e.g. Antonietti, 1991; Giaquinto, 2008; Stenning and Oberlander, 1995). I do not claim that diagrams can be used as proofs in mathematics. As is well known, we can be easily misled by accidental properties of diagrams. The concept of mathematical proof is different from the concept of mathematical construction.
11 As Black once said (1972, 122): 'My chief objection to the resemblance view, then, is that when pursued it turns out to be uninformative, offering a trivial verbal substitution in place of insight.'
12 This structure-preserving mapping does not have to be isomorphism. It can be either homomorphism (if the mapping is not bijective), monomorphism (if it is an injective homomorphism), epimorphism (if it is surjective homomorphism), endomorphism or automorphism (e.g. Pero and Suárez, 2016; Pietarinen, 2014). Determining the character of these mappings is irrelevant for my argument.
13 Frigg (2006, 50) introduces the so-called problem of style. It concerns the taxonomy of different styles of scientific representations. The riddle of style is similar to the problem of the multiple models, which can be expressed by the question of why scientists use different models of the same targets (Giere, 2006).

Chapter 5

1 Goodman holds (1976) that the idea of naturalness of icons can be explained by the concept of entrenchment. In a nutshell, icons are natural representations for we are more familiar with the representational systems that have a longer history of use. However, that cannot work as an explanation, for we can ask what properties of the system make it more likely to be entrenched and reproduced in history.
2 Although many philosophers of perception assume that we can perceive states of affairs (e.g. Armstrong, 1997; Fish, 2009; McDowell, 1996), it is not uncontested (e.g. Vernazzani, 2021). For our purposes, it is sufficient to claim that if the perception of facts exists, it is distinct from perceptual recognition.

3 In contrast to Gauker (2011), I do not hold that iconic representations cannot represent particulars as belonging to kinds. Gauker's claim seems to be challenged by the phenomenon of categorical perception. It refers to the common fact that we perceive distinct categories when there is a gradual change in a variable along a continuum. For instance, when we see a rainbow we experience colour stripes despite the stimulus is a smooth gradient of electromagnetic frequencies.

4 Much depends on how we think about the nature of memory. In contrast to the causal theories (e.g. Bernecker, 2010), the simulation theories (e.g. Michaelian, 2016) hold that memory draws on information acquired during past experience to construct a simulation of a target event. Memory is understood as a construction system, rather than a system based on retrieving information. However, regardless of how we think about the nature of memory, the causal story of how recognition works seems to be wrong.

5 This general description of recognition-based identification does not determine the psychological mechanisms of how such invariants are recognized. It can be reconciled with both view- and structure-based approaches to object recognition in cognitive psychology (e.g. Hummel, 2013).

6 Of course, it says little about how Cezanne achieves this goal in practice, for example, how the properties of the picture's surface, such as brush strokes in a painting, can produce the *experience* of seeing-in depicted colours. However, I strongly doubt that philosophers can answer this question. In the same way, I doubt that philosophers are in a position to answer the question of how strings of letters on a piece of paper are transformed into meaningful units by a cognitive system. Philosophers can ask about the nature of the meaning of language, not about the psychological mechanisms responsible for grasping the meaning. That is the job of psycholinguists. By the same token, a philosopher cannot answer the question of what construction invariants are for representing objects in practice.

Chapter 6

1 Greenberg (2018, 2021), following Cummins, introduces similar distinction between the target and content. According to Greenberg's Three-Part Model, the referent is a particular individual which the representation is of identified by the image singular content. Singular content is contrasted with the attributive content which identifies the properties ascribed to those individuals, for example their shape or orientation. The 2-D model differs from Three-Part Model in at least three respects. First, according to the 2-D model, a referent does not have to be a particular individual. Content can identify kinds. Second, the properties identified by the pictorial content are the construction properties. For the reasons mentioned

later, I hold that images lack predicative functions. Third, I hold that iconic content cannot be descriptive.
2. This distinction has been already applied to depiction in phenomenological tradition (e.g. Husserl, 2005; Ingarden, 1989; Wiesing, 2009). Husserl distinguishes between the picture's surface, its content (*Bildobjekt*) and target (*Bildsujet*).
3. Sometimes denotation is interpreted as restricted to linguistic representations, such as proper names. However, this restriction is neither necessary nor helpful (Elgin, 2010b).
4. To deal with targetless representations, Goodman (1976) distinguishes between representations-of and representations-as. A picture of a unicorn is not a representation of a unicorn but a kind of a picture, namely, a unicorn-picture. Kinds of pictures are distinguished by different modes of presentation. I do not endorse this distinction, for it reintroduces the old problem of sense-reference distinction. In the same way as there can be no mode of presentation without something that is represented (Evans, 1982), there can be no representation-as without representation-of. The same objection applies to Hyman's (2012, 2015) distinction between the sense and reference of pictorial representations.
5. Notably, we have to remember to keep aside representing a particular object by iconic content and identifying the possible world we are in. I hold that iconic content can identify particulars but is not relativized to the context of possible worlds (target is).
6. It is important not to confuse the understandability of images with their interpretability. Understandability of images refers to their ability to be used in some contexts. Interpretability abstracts from the context of use and determines the conditions that have to be satisfied by an image to bear a representational function. An image can be interpretable and not understandable. Suppose that in some population there are no mathematicians, and there is no one who can understand a mathematical graph. Still, the graph is interpretable. It does not work the other way round. An image cannot be understandable if it is not interpretable.
7. This description is complementary with the representational theory of measurement (RTM). According to RTM, measurement is 'the construction of homomorphisms (scales) from empirical relational structures of interest to numerical structures that are useful' (Krantz et al., 1971, 9). Measurement should satisfy two kinds of theorems. The representation theorem holds that there a represented domain can be mapped onto the structure of the representing domain. The uniqueness theorem establishes the permissible transformations of the represented domain into the numerical domain. However, our description of measurement does not depend on RTM.
8. Thinking about images in terms of measurement devices seems to imply that images possess arbitrary semantics. The argument can look as follows. If we can measure the world in any way we like, then we can depict the world any way we want, which is false. Images are naturally connected to the world. We cannot

depict horses as cows, whereas we can conduct measurements with many different measures. We can measure the length in metres and inches. However, this argument is invalid, for it is not true that measurements are arbitrarily connected to the world. We have different measurement conventions in the same way as there are different iconic conventions. Yet, it is not true that there is no difference in what type of measurement we choose. We can try to measure distances with gas, but it does not have the properties that enable identifying the distance. Thus, the way we measure depends on how the world is. In the same way, images are measurement devices that are naturally connected to the world. We can depict objects differently. However, the depiction has to possess properties that identify the object of depiction. Thus, images depend on what they represent.
9. The extensions of the Reutersvaard-Penrose triangle are the well-known Penrose stairs or Escher's 'Waterfall'.
10. Indeterminable constructions have to be distinguished from indetermined constructions. For instance, a picture of a man in a hat does not determine whether he is bald. However, it is a determinable property of this class of constructions. We can have a different picture of the same man without a hat.
11. The impossible images phenomenon is not restricted to spatial representations. Indeterminable constructions can be instantiated by temporal representations. For instance, echolocation jamming in bats occurs when non-target sounds interfere with target echoes and can be caused by the echolocation system itself.

Chapter 7

1. Note that the concept of understating comes in degrees. One does not have to understand a whole theory to understand its parts. Most people understand basic mathematical calculus, but it does not imply that they understand the advanced calculus.
2. This does not imply that division of language into meaningful parts is semantically non-arbitrary (Johnson, 2004; Quine, 1970; Salje, 2019). Yet, it is believed that language has a logical structure and a syntax.
3. Burge (2018) argues that we can establish canonical decomposition by discovering perceptual competences that determine picture constituents. By the same token, we can establish a picture's grammar-like structure. However, his argument misses the point, for Fodor holds that there is no principal way we can decompose a picture into semantical parts. Fodor's argument is metaphysical, not psychological.
4. In Kozak (2021), I distinguish between the so-called vehicle-specificity and modality-specificity of images.

5 It appears that it would be hard to understand the concept of mental imagery if it had nothing in common with the concept of images (e.g. Anderson, 1978).
6 For the discussion on the significance of dissociations between neural mechanisms of mental imagery and perception, see Bartolomeo et al. (2020) and Pearson (2020).
7 One can ask what mental images taken as conscious experiences are. That is an important question which, sadly, I cannot answer. It would require having a good theory of consciousness. This book is not about consciousness.
8 Amodal completion is not a visual phenomenon. It ranges over all modalities. For instance, we amodally complete a beeped part of a soundtrack (Nanay, 2018).

Literature

Aasen, S. (2016). 'Pictures, Presence and Visibility'. *Philosophical Studies*, 173 (1): 187–203.

Abel, G. (2012). 'Knowledge Research: Extending and Revising Epistemology'. In A. Abel and J. Conant (eds), *Rethinking Epistemology*, Vol. 1, 1–52. Berlin: DeGruyter.

Abell, C. (2009). 'Canny Resemblance: An Account of Depiction'. *Philosophical Review*, 118 (2): 183–223.

Abell, C. and K. Bantinaki (2010). 'Introduction'. In C. Abell and K. Bantinaki (eds), *Philosophical Perspectives on Depiction*, 1–24. Oxford; New York: Oxford University Press.

Abell, C. and G. Currie (1999). 'Internal and External Pictures'. *Philosophical Psychology*, 12 (4): 429–45.

Abid, G. (2021). 'Recognition and the Perception–Cognition Divide'. *Mind & Language*, 1–20. https://doi.org/10.1111/mila.12362.

Abrahamsen, A. and W. Bechtel (2015). 'Diagrams as Tools for Scientific Reasoning'. *Review of Philosophy and Psychology*, 6: 117–31.

Ambrosio, C. (2014). 'Iconic Representations and Representative Practices'. *International Studies in the Philosophy of Science*, 28 (3): 255–75.

Amit, E., C. Hoeflin, N. Hamzah and E. Fedorenko (2017). 'An Asymmetrical Relationship Between Verbal and Visual Thinking: Converging Evidence from Behavior and fMRI'. *Neuroimage*, 15, 619–27.

Anderson, J. R. (1978). 'Arguments Concerning Representations for Mental Imagery'. *Psychological Review*, 85 (4): 249–77.

Antonietti, A. (1991). 'Why Does Mental Visualisation Facilitate Problem-Solving?' In R. H. Logie and M. Denis (eds), *Mental Images in Human Cognition*, 211–27. Amsterdam: Elsevier.

Armstrong, D. (1968). *A Materialist Theory of the Mind*. London: Routledge & Kegan Paul.

Armstrong, D. (1989). *A Combinatorial Theory of Possibility*. Cambridge, MA: Cambridge University Press.

Armstrong, D. (1997). *A World of States of Affairs*. Cambridge, MA: Cambridge University Press.

Arnheim, R. (1969). *Visual Thinking*. Berkeley; Los Angeles: University of California Press.

Arnheim, R. (1980). 'A Plea for Visual Thinking'. *Critical Inquiry*, 6 (3): 489–97.

Ayers, M. (1991). *Locke: Epistemology and Ontology*. London: Routledge.

Baddeley, A. D. (1988). 'Imagery and Working Memory'. In M. Denis, J. Engelkamp and J. T. E. Richardson (eds), *Cognitive and Neuropsychological Approaches to Mental Imagery*, 169–80. Dordrecht: Martinus Nijhoff.

Baddeley, A. D. and K. Lieberman (1980). 'Spatial Working Memory'. In R. S. Nickerson (ed.), *Attention and Performance*, 521–39. Hillsdale, NJ: Erlbaum.

Baillargeon, R., E. Spelke and S. Wasserman (1985). 'Object Permanence in Five-Month-Old Infants'. *Cognition*, 20: 191–208.

Ball, D. N. (2018). 'Semantics as Measurement'. In D. Ball and B. Rabern (eds), *The Science of Meaning: Essays on the Metatheory of Natural Language Semantics*, 381–410. Oxford: Oxford University Press.

Bantinaki, K. (2007). 'Pictorial Perception as Illusion'. *British Journal of Aesthetics*, 47 (3): 268–79.

Barsalou, L. W. (1999). 'Perceptual Symbol Systems (With Commentaries and Author's Reply)'. *Behavioral and Brain Sciences*, 22: 577–660.

Barsalou, L. W. (2010). 'Grounded Cognition: Past, Present, and Future'. *Topics in Cognitive Science*, 2: 716–24.

Bartolomeo, P., D. Hajhajate, J. Liu and A. Spagna (2020). 'Assessing the Causal Role of Early Visual Areas in Visual Mental Imagery'. *Nature Reviews Neuroscience*, 21: 517.

Barwise, J. and J. Etchemendy (1996). 'Visual Information and Valid Reasoning'. In G. Allwein and J. Barwise (eds), *Logical Reasoning with Diagrams*, 3–26. Oxford: Oxford University Press.

Barwise, J. and A. Shimojima (1995). 'Surrogate Reasoning'. *Cognitive Studies: Bulletin of Japanese Cognitive Science Society*, 4 (2): 7–27.

Bauer, M. I. and P. N. Johnson-Laird (1993). 'How Diagrams Can Improve Reasoning'. *Psychological Science*, 4, 372–8.

Bayne, T. (2013). *Thought: A Very Short Introduction*. Oxford: Oxford University Press.

Beauchamp, M. S. and A. Martin (2007). 'Grounding Object Concepts in Perception and Action: Evidence from fMRI Studies of Tools'. *Cortex*, 43 (3): 461–8.

Bechtel, W. (2008). *Mental Mechanisms: Philosophical Perspectives on Cognitive Neuroscience*. New York: Lawrence Erlbaum Associates.

Bechtel, W. (2017). 'Diagrammatic Reasoning'. In L. Magnani and T. Bertolotti (eds), *Handbook of Model-Based Science*, 605–18. Dordrecht: Springer.

Beck, J. (2015). 'Analogue Magnitude Representations: A Philosophical Introduction'. *The British Journal for the Philosophy of Science*, 66 (4): 829–55.

Bedny, M., A. Caramazza, E. Grossman, A. Pascual-Leone and R. Saxe (2008). 'Concepts Are More Than Percepts: The Case of Action Verbs'. *The Journal of Neuroscience*, 28: 11347–53.

Bedny, M., A. Caramazza, A. Pascual-Leone and R. Saxe (2012). 'Typical Neural Representations of Action Verbs Develop Without Vision'. *Cerebral Cortex*, 22: 286–93.

Beilock, S. L. and S. Goldin-Meadow (2010). 'Gesture Changes Thought by Grounding It in Action'. *Psychological Science*, 21: 1605–10.

Belting, H. (2001). *Bild-Anthropologie. Entwürfe für eine Bildwissenschaft*. München: Wilhelm Fink Verlag.
Ben-Yami, H. (1997). 'Against Characterizing Mental States as Propositional Attitudes'. *The Philosophical Quarterly*, 47: 84–9.
Bennett, M. and P. Hacker (2003). *Philosophical Foundations of Neuroscience*. Oxford: Blackwell.
Bensafi, M., J. Porter, S. Pouliot, J. Mainland, B. Johnson, C. Zelano, N. Young, E. Bremner, D. Aframian, R. Kahn and N. Sobel (2003). 'Olfactomotor Activity During Imagery Mimics that During Perception'. *Nature Neuroscience*, 6 (11): 1142–4.
Bensafi, M., S. Pouliot and N. Sobel (2005). 'Odorant-Specific Patterns of Sniffing During Imagery Distinguish "Bad" and "Good" Olfactory Imagers'. *Chemical Senses*, 30 (6): 521–9.
Berman, D. (2008). 'A Confession of Images'. *Philosophical Practice*, 3: 255–66.
Bermúdez, J. L. (2000). 'Naturalized Sense-Data'. *Philosophy and Phenomenal Research*, 61, 353–74.
Bermúdez, J. L. (2003). *Thinking Without Words*. Oxford: Oxford University Press.
Bernecker, S. (2010). *Memory: A Philosophical Study*. Oxford: Oxford University Press.
Black, M. (1972). 'How Do Pictures Represent?' In E. H. Gombrich, J. Hochberg, M. Black (eds), *Art, Perception, and Reality*, 95–130. Baltimore; London: The John Hopkins UP.
Blackwell, S. E. (2020). 'Emotional Mental Imagery'. In A. Abraham (ed.), *The Cambridge Handbook of the Imagination*, 241–57. Cambridge: Cambridge University Press.
Block, N. (1983a). 'Mental Pictures and Cognitive Science'. *The Philosophical Review*, 92 (4): 499–541.
Block, N. (1983b). 'The Photographic Fallacy in the Debate About Mental Imagery'. *Noûs*, 17 (4): 651–62.
Blumson, B. (2009). 'Defining Depiction'. *British Journal of Aesthetics*, 49 (2): 143–57.
Blumson, B. (2012). 'Mental Maps'. *Philosophy and Phenomenological Research*, 85 (2): 413–34.
Blumson, B. (2014). *Resemblance and Representation: An Essay in the Philosophy of Pictures*. Cambridge: Open Book Publishers.
Bocanegra, B. R., F. H. Poletiek, B. Ftitache and A. Clark (2019). 'Intelligent Problem-Solvers Externalize Cognitive Operations'. *Nature Human Behaviour*, 3: 136–42.
Boehm, G. (2007). *Wie Bilder Sinn erzeugen: Die Macht des Zeigens*. Berlin: Berlin University Press.
Bordwell, D. (2008). *Poetics of Cinema*. New York: Routledge.
Botterill, G. and P. Carruthers (1999). *The Philosophy of Psychology*. Cambridge: Cambridge University Press.
Boumans, M. (2005). *How Economists Model the World into Numbers*. New York: Routledge.

Bradburn, N. M., N. L. Cartwright and J. Fuller (2017). 'A Theory of Measurement'. In L. McClimans (ed.), *Measurement in Medicine: Philosophical Essays on Assessment and Evaluation*, 73–88. London: Rownan & Littlefield International.

Braddon-Mitchell, D. and F. Jackson (1996). *Philosophy of Mind and Cognition*. Oxford: Blackwell.

Braine, M. D. S. and D. P. O'Brien (1998). 'The Theory of Mental-Propositional Logic: Description and Illustration'. In M. D. S. Braine and D. P. O'Brien (eds), *Mental Logic*, 79–89. Mahwah, NJ: Lawrence Erlbaum Associates.

Briscoe, R. (2016). 'Depiction, Pictorial Experience, and Vision Science'. *Philosophical Topics*, 44 (2): 43–81.

Brown, J. R. (1999). *Philosophy of Mathematics: An Introduction to the World of Proofs and Pictures*. London: Routledge.

Buccella, A. (2021). 'The Problem of Perceptual Invariance'. *Synthese*, 199 (5–6): 13883–905.

Budd, M. (1995). *Values of Art*. London: Allen Lane.

Budd, M. (2008). *Aesthetic Essays*. Oxford: Oxford University Press.

Bueno, O. and S. French (2011). 'How Theories Represent'. *The British Journal for the Philosophy of Science*, 62 (4): 857–94.

Burge, T. (2010). *The Origins of Objectivity*. Oxford: Oxford University Press.

Burge, T. (2018). 'Iconic Representation: Maps, Pictures, and Perception'. In S. Wuppuluri and F. A. Doria (eds), *The Map and the Territory: Exploring the Foundations of Science, Thought and Reality*, 79–100. Cham, Switzerland: Springer.

Burke, L. (1952). 'On the Tunnel Effect'. *Quarterly Journal of Experimental Psychology*, 4: 121–38.

Byrne, R. M. J. and P. N. Johnson-Laird (1989). 'Spatial Reasoning'. *Journal of Memory and Language*, 28: 564–75.

Camp, E. (2004). 'The Generality Constraint and Categorial Restrictions'. *Philosophical Quarterly*, 54 (215): 209–31.

Camp, E. (2007). 'Thinking with Maps'. In J. Hawthorne (ed.), *Philosophical Perspectives 21: Philosophy of Mind*, 145–82. Oxford: Wiley-Blackwell.

Camp, E. (2015). 'Logical Concepts and Associative Characterizations'. In E. Margolis and S. Laurence (eds), *The Conceptual Mind: New Directions in the Study of Concepts*, 591–621. Cambridge, MA: MIT Press.

Camp, E. (2018). 'Why Maps Are Not Propositional'. In A. Grzankowski and M. Montague (eds), *Non-Propositional Intentionality*, 19–45. Oxford: Oxford University Press.

Candlish, S. (2001). 'Mental Imagery'. In S. Schroeder (ed.), *Wittgenstein and Contemporary Philosophy of Mind*, 107–28. London; New York: Palgrave MacMillan.

Cantlon, J. F. (2012). 'Math, Monkeys, and the Developing Brain'. *Proceedings of the National Academy of Sciences*, 109 (Supplement 1): 10725–32.

Carey, S. (2009). *The Origin of Concepts*. New York: Oxford University Press.

Carey, S. (2011). 'Précis of "The Origin of Concepts"'. *The Behavioral and brain sciences*, 34 (3): 113–24.

Cartwright, N. (1983). *How the Laws of Physics Lie*. Oxford: Oxford University Press.

Cartwright, N. and L. Runhardt (2014). 'Measurement'. In N. Cartwright and E. Montuschi (eds), *Philosophy of Social Science: A New Introduction*, 265–87. Oxford: Oxford University Press.

Carvalho, J. M. (2019). *Thinking with Images: An Enactivist Aesthetics*. New York; London: Routledge.

Casati, R. (2015). 'Object Perception'. In M. Matthen (ed.), *The Oxford Handbook of Philosophy of Perception*, 393–404. Oxford: Oxford University Press.

Casati, R. (2018). 'Two, then Four Modes of Functioning of the Mind: Towards a Unification of "Dual" Theories of Reasoning and Theories of Cognitive Artefacts'. In J. M. Zacks and H. A. Taylor (eds), *Representations in Mind and World: Essays inspired by Barbara Tversky*, 7–23. New York: Routledge.

Casati, R. and A. Varzi (1999). *Parts and Places: The Structures of Spatial Representation*. Cambridge, MA: MIT Press.

Cavedon-Taylor, D. (2021). 'Untying the Knot: Imagination, Perception and Their Neural Substrates'. *Synthese*, 199: 7203–30.

Chambers, D. and D. Reisberg (1985). 'Can Mental Images be Ambiguous?' *Journal of Experimental Psychology: Human Perception and Performance*, 11 (3): 317–28.

Chang, H. (2004). *Inventing Temperature: Measurement and Scientific Progress*. Oxford: Oxford University Press.

Chang, H. and N. Cartwright (2014). 'Measurement'. In M. Curd and S. Psillos (eds), *The Routledge Companion to Philosophy of Science*, 411–19. London; New York: Routledge.

Chappell, V. (1994). 'Locke's Theory of Ideas'. In V. Chappell (ed.), *The Cambridge Companion to Locke*, 26–55. Cambridge: Cambridge University Press.

Chevalier, J.-M. (2015). 'The Problem of Resemblance in Peirce's Semiotics and Philosophy'. *Versus*, 120: 45–59.

Church, J. (2008). 'The Hidden Image: A Defense of Unconscious Imagining and Its Importance'. *American Imago*, 65: 379–404.

Clark, A. (2013). 'Whatever Next? Predictive Brains, Situated Agents, and the Future of Cognitive Science'. *Behavioral and Brain Sciences*, 36: 181–204.

Clement, C. A. and R. J. Falmagne (1986). 'Logical Reasoning, World Knowledge, and Mental Imagery: Interconnections in Cognitive Processes'. *Memory and Cognition*, 14: 299–307.

Collins, C. (1991). *The Poetics of the Mind's Eye: Literature and the Psychology of Imagination*. Philadelphia, PA: University of Pennsylvania Press.

Coopmans, C. (2014). 'Visual Analytics as Artful Revelation'. In C. Coopmans, J. Vertesi, M. Lynch and S. Woolgar (eds), *Representation in Scientific Practice Revisited*, 37–60. Cambridge, MA; London: MIT Press.

Crane, T. (1988). 'The Waterfall Illusion'. *Analysis*, 48: 142–7.

Crane, T. (2001). *Elements of Mind*. Oxford: Oxford University Press.
Crane, T. (2009). 'Is Perception a Propositional Attitude?' *The Philosophical Quarterly*, 59: 452–69.
Craven, J. and D. H. Foster (1992). 'An Operational Approach to Colour Constancy'. *Vision Research*, 32 (7): 1359–66.
Croijmans, I., L. Speed, A. Arshamian and A. Majid (2019). 'Measuring the Multisensory Imagery of Wine: The Vividness of Wine Imagery Questionnaire'. *Multisensory Research*, 32 (3): 179–95.
Cromwell, P. (2004). *Knots and Links*. Cambridge: Cambridge University Press.
Cummins, R. (1996). *Representations, Targets, and Attitudes*. Cambridge, MA; London: MIT Press.
Curry, G. (1995). *Image and Mind: Film, Philosophy and Cognitive Science*. New York: Cambridge University Press.
Davidson, D. (1982). 'Rational Animals'. *Dialectica*, 36 (4): 317–27.
Davidson, D. (1997). 'Seeing Through Language". In J. Preston (ed.), *Thought and Language*, 15–27. Cambridge: Cambridge University Press.
Davidson, D. (2001). *Inquiries into Truth and Interpretation*. Oxford: Oxford University Press.
Davies, W. (2016). 'Colour Constancy, Illumination, and Matching'. *Philosophy of Science*, 83: 540–62.
Dawes, A. J., R. Keogh, T. Andrillon and J. Pearson (2020). 'A Cognitive Profile of Multi-Sensory Imagery, Memory and Dreaming in Aphantasia'. *Scientific Reports*, 10 (1): 1–10.
Defeyter, M. A., R. Russo and P. L. McPartlin (2009). 'The Picture Superiority Effect in Recognition Memory: A Developmental Study Using the Response Signal Procedure'. *Cognitive Development*, 24 (3): 265–73.
Dehaene, S. (2011). *The Number Sense: How The Mind Creates Mathematics*. Oxford: Oxford University Press.
DeLoache, J., S. L. Pierroutsakos and D. H. Uttal (2003). 'The Origins of Pictorial Competence'. *Current Directions in Psychological Science*, 12 (4): 114–18.
Denis, M. (1991). 'Imagery and Thinking'. In C. Cornoldi and M. McCaniel (eds), *Imagery and Cognition*, 103–31. New York: Springer Verlag.
Dennett, D. C. (1981a). 'The Nature of Images and the Introspective Trap'. In N. Block (ed.), *Imagery*, 51–62. New York; London: MIT Press.
Dennett, D. C. (1981b). 'Two Approaches to Mental Images'. In. N. Block (ed.), *Imagery*, 87–107. New York; London: MIT Press.
Dennett, D. C. (1987). *The Intentional Stance*. Cambridge, MA: MIT Press.
Dennett, D. C. (1990). 'Thinking with a Computer'. In H. Barlow, C. Blakemore and M. Westo-Smith (eds), *Images and Understanding: Thoughts about Images: Ideas about Understanding*, 297–309. Cambridge: Cambridge University Press.
Dennett, D. C. (1991a). *Consciousness Explained*. Boston, MA: Little, Brown.
Dennett, D. C. (1991b). 'Real Patterns'. *Journal of Philosophy*, 88 (1): 27–51.

Dennett, D. C. (2013). *Intuition Pumps and Other Tools for Thinking*. New York: W. W. Norton & Company.

Dennett, D. C. (2017). *From Bacteria to Bach and Back: The Evolution of Minds*. New York: W. W. Norton & Company.

De Regt, H. W. (2014). 'Visualization as a Tool for Understanding'. *Perspectives on Science*, 22 (3): 377–96.

De Regt, H. W. (2017). *Understanding Scientific Understanding*. New York: Oxford University Press.

De Toffoli, S. and V. Giardino (2014). 'Forms and Roles of Diagrams in Knot Theory'. *Erkenntnis*, 79 (4): 829–42.

Devitt, M. (2006). *Ignorance of Language*. Oxford: Oxford University Press.

Dilworth, J. (2008). 'The Propositional Challenge to Aesthetics'. *British Journal of Aesthetics*, 48 (2): 115–44.

Djordjevic, J., R. J. Zatorre, M. Petrides, J. A. Boyle and M. Jones-Gotman (2005). 'Functional Neuroimaging of Odor Imagery'. *Neuroimage*, 24: 791–801.

Djordjevic, J., R. J. Zatorre, M. Petrides and M. Jones-Gotman (2004). 'The Mind's Nose: Effects of Odor and Visual Imagery on Odor Detection'. *Psychological Science*, 15 (3): 143–8.

Dokic, J. (2010). 'Perceptual Recognition and the Feeling of Presence'. In B. Nanay (ed.), *Perceiving the World*, 33–53. Oxford: Oxford University Press.

Dominowski, R. L. and L. E. Bourne (1994). 'History of Research on Thinking and Problem Solving'. In R. J. Sternberg (ed.), *Thinking and Problem Solving*, 1–36. San Diego, CA: Academic Press.

Dorsch, F. (2016). 'Knowledge by Imagination – How Imaginative Experience Can Ground Factual Knowledge'. *Teorema: Revista Internacional De Filosofía*, 35 (3): 87–116.

Dresner, E. (2002). 'Measurement Theoretic Semantics and Semantics of Necessity'. *Synthese*, 130: 413–40.

Dresner, E. (2006). 'A Measurement Theoretic Account of Propositions'. *Synthese*, 153: 1–22.

Dresner, E. (2010). 'Language and the Measure of Mind'. *Mind & Language*, 25 (4): 418–39.

Dresner, E. (2014). 'Decision Theory, Propositional Measurement, and Unified Interpretation'. *Mind*, 123 (491): 707–32.

Dretske, F. (1981). *Knowledge and the Flow of Information*. Cambridge, MA: MIT Press.

Dummett, M. (1993). *Origins of Analytical Philosophy*. London: Duckworth.

Earnshaw, R. A. and D. Watson (eds) (1993). *Animation and Scientific Visualization: Tools and Applications*. London: Academic Press.

Eaton, M. (1980). 'Truth in Pictures'. *The Journal of Aesthetics and Art Criticism*, 39 (1): 15–26.

Eco, U. (1995). *The Search for the Perfect Language*. Oxford: Blackwell.

Egan, D. (2011). 'Pictures in Wittgenstein's Later Philosophy'. *Philosophical Investigations*, 34 (1): 55–76.

Elgin, C. Z. (1983). *With Reference to Reference*. Indianapolis; Cambridge: Hackett.

Elgin, C. Z. (1996a). *Considered Judgment*. Princeton, NJ: Princeton University Press.

Elgin, C. Z. (1996b). 'Index and Icon Revisited'. In V. M. Colapietro and T. M. Olshewsky (eds), *Peirce's Doctrine of Signs: Theory, Applications, and Connections*, 181–9. Berlin; New York: de Gruyter.

Elgin, C. Z. (2010a). 'Keeping Things in Perspective'. *Philosophical Studies*, 150: 439–47.

Elgin, C. Z. (2010b). 'Telling Instances'. In R. Frigg and M. C. Hunter (eds), *Beyond Mimesis and Convention: Representation in Art and Science*, 1–18. Berlin; New York: Springer.

Elgin, C. Z. (2017). *True Enough*. Cambridge, MA; London: MIT Press.

Ellis, R. D. (1995). *Questioning Consciousness: The Interplay of Imagery, Cognition, and Emotion in the Human Brain*. Amsterdam: John Benjamins.

Elpidorou, A. (2016). 'Seeing the Impossible'. *Journal of Aesthetics and Art Criticism*, 74 (1): 11–21.

Ernst, B. (1986). *Adventures with Impossible Figures*. Norfolk: Tarquin Publications.

Ertz, T.-P. (2008). *Regel und Witz: Wittgensteinsche Perspektiven auf Mathematik, Sprache und Moral*. Berlin; New York: de Gruyter.

Evans, G. (1982). *The Varieties of Reference*. Oxford: Oxford University Press.

Evans, J. St B. T. (2017). *Thinking and Reasoning: A Very Short Introduction*. Oxford: Oxford University Press.

Fazekas, P. and B. Nanay (2017). 'Pre-Cueing Effects: Attention or Mental Imagery'. *Frontiers in Psychology*, 8: 222.

Ferguson, E. S. (1992). *Engineering and the Mind's Eye*. Cambridge, MA: MIT Press.

Finke, R. A. (1989). *Principles of Mental Imagery*. Cambridge, MA: MIT Press.

Finke, R. A., S. Pinker and M. J. Farah (1989). 'Reinterpreting Visual Patterns in Mental Imagery'. *Cognitive Science*, 13 (1): 51–78.

Finkelstein, L. (2003). 'Widely, Strongly and Weakly Defined Measurement'. *Measurement*, 34: 39–48.

Firestone, C. and B. J. Scholl (2016). 'Cognition Does Not Affect Perception: Evaluating the Evidence for "Top-Down" Effects'. *Behavioral and Brain Sciences*, 36: 1–77.

Fish, W. (2009). *Perception, Hallucination, and Illusion*. New York: Oxford University Press.

Fodor, J. (1975). *The Language of Thought*. Cambridge, MA: Harvard University Press.

Fodor, J. (1987). *Psychosemantics: The Problem of Meaning in the Philosophy of Mind*. Cambridge, MA: MIT Press.

Fodor, J. (1991). 'Replies'. In B. Loewer and G. Rey (eds), *Meaning in Mind: Fodor and his Critics*, 255–319. Oxford: Basil Blackwell.

Fodor, J. (2007). 'The Revenge of the Given'. In B. P. McLaughlin and J. D. Cohen (eds), *Contemporary Debates in Philosophy of Mind*, 105–16. Oxford: Basil Blackwell.

Fodor, J. (2008). *LOT 2: The Language of Thought Revisited*. Oxford: Oxford University Press.

Fodor, J. and Z. Pylyshyn (1981). 'How Direct is Visual Perception?: Some Reflections on Gibson's "Ecological Approach"'. *Cognition*, 9: 207–46.

Fodor, J. and Z. Pylyshyn (1988). 'Connectionism and Cognitive Architecture: A Critical Analysis'. *Cognition*, 28 (1–2): 3–71.

Foster, D. (2003). 'Does Colour Constancy Exist?' *Trends in Cognitive Sciences*, 7: 439–43.

Foster, D. (2011). 'Colour Constancy'. *Vision Research*, 51: 674–700.

Francis, G. K. (2007). *A Topological Picturebook*. New York: Springer Verlag.

Frege, G. (1984). 'Thoughts'. Translated by P. Geach and R. Stoothoff. In B. McGuinness (ed.), *Collected Papers on Mathematics, Logic, and Philosophy*, 351–72. Oxford: Basil Blackwell.

French, S. R. D. (2003). 'A Model-Theoretic Account of Representation (Or, I Don't Know Much About Art But I Know It Involves Isomorphism)'. *Philosophy of Science*, 70 (5): 1472–83.

Frigg, R. (2006). 'Scientific Representation and the Semantic View of Theories'. *Theoria*, 21 (1): 49–65.

Frigg, R and J. Nguyen (2020). *Modelling Nature: An Opinionated Introduction to Scientific Representation*. Cham: Springer.

Frixione, M. and A. Lombardii (2015). 'Street Signs and Ikea Instruction Sheets: Pragmatics and Pictorial Communication'. *Review of Philosophy and Psychology*, 6: 133–49.

Galton, F. (1880). 'Statistics of Mental Imagery'. *Mind*, 5: 301–18.

Gardner, H. (2004). *Frames of Mind: The Theory of Multiple Intelligences*. New York: Basic Books.

Gaskin, R. (2008). *The Unity of the Proposition*. Oxford: Oxford University Press.

Gauker, C. (2007). 'A Critique of the Similarity Space Theory of Concepts'. *Mind & Language*, 22 (4): 317–45.

Gauker, C. (2011). *Words and Images: An Essay on the Origin of Ideas*. Oxford: Oxford university Press.

Gauker, C. (2017). 'Three Kinds of Nonconceptual Seeing-as'. *Review of Philosophy and Psychology*, 8 (4): 763–79.

Gauker, C. (2020). 'On the Difference Between Realistic and Fantastic Imagining'. *Erkenntnis*, 1–20.

Gendler, T. S. (2004). 'Thought Experiments Rethought – and Reperceived'. *Philosophy of Science*, 71: 1154–63.

Gersel, J. P. (2017). 'Discriminatory Capacities, Russell's Principle, and the Importance of Losing Sight of Objects'. *European Journal of Philosophy*, 25: 700–20.

Giaquinto, M. (2007). *Visual Thinking in Mathematics: An Epistemological Study*. Oxford: Oxford University Press.

Giaquinto, M. (2008). 'Visualizing in Mathematics'. In P. Mancosu (ed.), *The Philosophy of Mathematical Practice*, 22–42. Oxford: Oxford University Press.

Giaquinto, M. (2011). 'Crossing Curves: A Limit to the Use of Diagrams in Proofs'. *Philosophia Mathematica*, 19: 281–307.
Giardino, V. (2014). 'Diagramming: Connecting Cognitive Systems to Improve Reasoning'. In A. Benedek and K. Nyiri (eds), *The Power of the Image: Emotion, Expression, Explanation*, 23–34. Frankfurt/M: Peter Lang.
Giardino, V. (2016). 'Behind the Diagrams: Cognitive Issues and Open Problems'. In S. Krämer and C. Ljungberg (eds), *Thinking with Diagrams: The Semiotic Basis of Human Cognition*, 77–101. Boston; Berlin: Walter de Gruyter.
Giardino, V. (2017). 'Diagrammatic Reasoning in Mathematics'. In L. Magnani and T. Bertolotti (eds), *Handbook of Model-Based Science*, 499–522. Dordrecht: Springer.
Giardino, V. and G. Greenberg (2015). 'Introduction: Varieties of Iconicity'. *Review of Philosophy and Psychology*, 6 (1): 1–25.
Gibson, J. J. (1954). 'A Theory of Pictorial Perception'. *Audio Visual Communication Review*, 2 (1): 3–23.
Gibson, J. J. (1971). 'The Information Available in Pictures'. *Leonardo*, 4: 27–35.
Gibson, J. J. (1973). 'On the Concept of "Formless Invariants" in Visual Perception'. *Leonardo*, 6: 43–5.
Gibson, J. J. (1978). 'The Ecological Approach to the Visual Perception of Pictures'. *Leonardo*, 11: 227–35.
Gibson, J. J. (1979). *The Ecological Approach to the Visual Perception*. Houghton, MI: Mifflin.
Giere, R. N. (2006). *Scientific Perspectivism*. Chicago; London: University of Chicago Press.
Gilbert, A. N., M. Crouch and S. E. Kemp (1998). 'Olfactory and Visual Mental Imagery'. *Journal of Mental Imagery*, 22: 137–46.
Giovannelli, A. (2001). 'Picture, Image and Experience: A Philosophical Inquiry'. *Mind*, 110: 481–5.
Goldberg, B. (2017). 'Mechanism and Meaning'. In J. Hyman (ed.), *Investigating Psychology: Sciences of the Mind After Wittgenstein*, 48–66. London, New York: Routledge.
Goldin-Meadow, S. (2003). *Hearing Gesture: How Our Hands Help Us to Think*. Cambridge, MA: Harvard University Press.
Goldstone, R. L. and L. W. Barsalou (1998). 'Reuniting Perception and Conception'. *Cognition*, 65: 231–62.
Gombrich, E. H. (1960). *Art and Illusion: A Study in the Psychology of Pictorial Representation*. London: Phaidon.
Gombrich, E. H. (1971). 'Exchange of Letters'. *Leonardo*, 4: 195–203.
Gombrich, E. H. (1972). 'The "What" and the "How": Perspective Representation and the Phenomenal World'. In R. S. Rudner and I. Scheffler (eds), *Logic and Art: Essays in Honor of Nelson Goodman*, 129–49. Indianapolis, IN: Bobbs-Merrill.

Gombrich, E. H. (1990). 'Pictorial Instructions'. In H. Barlow, C. Blakemore and M. Weston-Smith (eds), *Images and Understanding*, 26–45. Cambridge, MA: Cambridge University Press.

Gooding, D. C. (1990). *Experiment and the Making of Meaning: Human Agency in Scientific Observation and Experiment*. Dordrecht; Boston: Kluwer Academic Publishers.

Gooding, D. C. (2010). 'Visualizing Scientific Inference'. *Topics in Cognitive Science*, 2: 15–35.

Goodman, N. (1972). 'Seven Strictures on Similarity'. In N. Goodman (ed.), *Problems and Projects*, 437–46. Indianapolis; New York: The Bobbs-Merrill Company.

Goodman, N. (1976). *Languages of Art*, 2nd edn. Indianapolis, IN: Hackett.

Goodman, N. (1984). *Of Mind and Other Matters*. Cambridge, MA; London: Harvard University Press.

Goodman, N. (1988). 'On What Should Not Be Said about Representation'. *Journal of Aesthetics and Art Criticism*, 46 (3): 419.

Goodman, N. (1990). 'Pictures in the Mind?' In H. Barlow, C. Blakemore and M. Westo-Smith (eds), *Images and Understanding: Thoughts about Images: Ideas about Understanding*, 358–64. Cambridge: Cambridge University Press.

Goodman, N. and C. Z. Elgin (1988). *Reconceptions in Philosophy and other Arts and Sciences*. Indianapolis; Cambridge: Hackett.

Goodwin, W. (2009). 'Visual Representation in Science'. *Philosophy of Science*, 76: 372–90.

Grandin, T. (2006). *Thinking in Pictures and Other Reports from My Life with Autism*. London: Bloomsbury.

Green, E. J. (2019). 'On the Perception of Structure'. *Noûs*, 53 (3): 564–92.

Green, E. J. (2020). 'The Perception-Cognition Border: A Case for Architectural Division'. *Philosophical Review*, 129: 323–93.

Green, E. J. and J. Quilty-Dunn (2021). 'What is an Object File?' *The British Journal for the Philosophy of Science*, 72 (3): 665–99.

Greenberg, D. L., M. J. Eacott, D. Brechin and D. C. Rubin (2005). 'Visual Memory Loss and Autobiographical Amnesia: A Case Study'. *Neuropsychologia*, 43 (10): 1493–502.

Greenberg, G. (2013). 'Beyond Resemblance'. *Philosophical Review*, 122 (2): 215–87.

Greenberg, G. (2018). 'Content and Target in Pictorial Representation'. *Ergo*, 5 (33): 865–98.

Greenberg, G. (2021). 'Semantics of Pictorial Space'. *Review of Philosophy and Psychology*, 12: 847–87.

Gregory, D. (2013). *Showing, Sensing, and Seeming: Distinctively Sensory Representations and their Contents*. Oxford: Oxford University Press.

Gregory, D. (2020). 'Pictures, Propositions, and Predicates'. *American Philosophical Quarterly*, 57 (2): 155–70.

Gregory, R. L. (1966). *Eye and Brain*. London: World University Library.

Grzankowski, A. (2013). 'Non-Propositional Attitudes'. *Philosophy Compass*, 8 (12): 1123–37.

Grzankowski, A. (2015). 'Pictures Have Propositional Content'. *Review of Philosophy and Psychology*, 6: 151–63.

Grzankowski, A. (2018). 'A Theory of Non-Propositional Attitudes'. In A. Grzankowski and M. Montague (eds), *Non-Propositional Intentionality*, 134–51. Oxford: Oxford University Press.

Guillot, A. (2020). 'Neurophysiological Foundations and Practical Applications of Motor Imagery'. In A. Abraham (eds), *The Cambridge Handbook of the Imagination*, 207–26. Cambridge: Cambridge University Press.

Gurr, C., J. Lee and K. Stenning (1998). 'Theories of Diagrammatic Reasoning: Distinguishing Component Problems'. *Minds and Machines*, 8: 533–57.

Halpern, A. R., R. J. Zatorre, M. Bouffard and J. A. Johnson (2004). 'Behavioral and Neural Correlates of Perceived and Imagined Musical Timbre'. *Neuropsychologia*, 42 (9): 1281–92.

Haugeland, J. (1998). *Having Thought: Essays in the Metaphysics of Mind*. Cambridge, MA; London: Harvard University Press.

Hawthorne, J. and D. Manley (2012). *The Reference Book*. Oxford: Oxford University Press.

Heck, R. G. (2000). 'Nonconceptual Content and the "Space of Reasons"'. *The Philosophical Review*, 109: 483–523.

Heck, R. G. (2007). 'Are There Different Kinds of Content?' In B. P. McLaughlin and J. Cohen (eds), *Contemporary Debates in Philosophy of Mind*, 117–38. Oxford: Blackwell.

Hegarty, M. (1992). 'Mental Animation: Inferring Motion From Static Displays of Mechanical Systems'. *Journal of Experimental Psychology: Learning, Memory, and Cognition*, 18: 1084–102.

Hegarty, M. and M. Kozhevnikov (1999). 'Types of Visual-Spatial Representations and Mathematical Problem Solving'. *Journal of Educational Psychology*, 91 (4): 684–9.

Hintikka, J. (1987). 'Mental Models, Semantical Games, and Varieties of Intelligence'. In L. Vaina (ed.), *Matters of Intelligence: Conceptual Structures in Cognitive Neuroscience*, 197–215. Dordrecht: D. Reidel.

Hinton, G. (1979). 'Some Demonstrations of the Effects of Structural Descriptions in Mental Imagery'. *Cognitive Science*, 3 (3): 231–50.

Holyoak, K. J. and R. G. Morrison (2012). 'Thinking and Reasoning: A Reader's Guide'. In K. J. Holyoak and R. G. Morrison (eds), *Oxford Handbook of Thinking and Reasoning*, 1–7. New York: Oxford University.

Holyoak, K. J. and P. Thagard (1996). *Mental Leaps: Analogy in Creative Thought*. Cambridge, MA; London: MIT Press.

Hookway, C. (2000). *Truth, Rationality and Pragmatism*. Oxford: Oxford University Press.

Hookway, C. (2007). 'Peirce on Icons and Cognition'. In U. Priss, S. Polovina and R. Hill (eds), *Conceptual Structures: Knowledge Architectures for Smart Applications*, 59–68. Berlin; Heidelberg: Springer.

Hope, V. (2009). 'Object Perception, Perceptual Recognition, and That-Perception Introduction'. *Philosophy*, 84 (330): 515–28.

Hopkins, R. (1998). *Picture, Image and Experience: A Philosophical Inquiry*. Cambridge: Cambridge University Press.

Hubbard, T. L. (1997). 'Target Size and Displacement Along the Axis of Implied Gravitational Attraction: Effects of Implied Weight and Evidence of Representational Gravity'. *Journal of Experimental Psychology: Learning, Memory, and Cognition*, 23: 1484–93.

Hubbard, T. L. (2010). 'Auditory Imagery: Empirical Findings'. *Psychological Bulletin*, 136 (2): 302–29.

Hume, D. (1975). *A Treatise of Human Nature*. Ed. L. A. Selby-Bigge, 2nd edn., revised by P. H. Nidditch. Oxford: Clarendon Press.

Hume, D. (1977). *An Enquiry Concerning Human Understanding*. Ed. E. Steinberg. Indianapolis, IN: Hackett.

Hummel, J. E. (2013). 'Object Recognition'. In D. Reisberg (ed.), *Oxford Handbook of Cognitive Psychology*, 32–46. Oxford: Oxford University Press.

Husserl, E. (2005). *Phantasy, Image Consciousness, and Memory (1898–1925)*. Trans. R. Bernet. Dordrecht: Springer.

Hyman, J. (2006). *The Objective Eye: Color, Form, and Reality in the Theory of Art*. Chicago, IL: University of Chicago Press.

Hyman, J. (2012). 'Depiction'. *Royal Institute of Philosophy Supplements*, 71: 129–50.

Hyman, J. (2015). 'Depiction'. In P. F. Bundgaard and F. Stjernfelt (eds), *Investigations in the Phenomenology and the Ontology of the Work of Art: What are Artworks and How Do We Experience Them?*, 191–208. Heidelberg: Springer.

Ingarden, R. (1989). *The Ontology of the Work of Art*. Trans. R. Meyer and J. T. Goldthwait. Athens, OH: Ohio University Press.

Inkpin, A. (2016). 'Projection, Recognition, and Pictorial Diversity'. *Theoria*, 82: 32–55.

International Organization for Standardization (1994). *Accuracy (Trueness and Precision) of Measurement Methods and Results – Part 1: General Principles and Definitions*. Standard No. 5725-1.

Intons-Peterson, M. J. (1983). 'Imagery Paradigms: How Vulnerable are They to Experimenter's Expectations?' *Journal of Experimental Psychology: Human Perception and Performance*, 9: 394–412.

Intons-Peterson, M. J. and A. R. White (1981). 'Experimenter Naiveté and Imaginal Judgments'. *Journal of Experimental Psychology: Human Perception and Performance*, 7 (4): 833–43.

Irvine, E. (2014). 'What Iconicity Can and Cannot Do for Proto-Language'. In E. A. Cartmill, S. Roberts, H. Lyn and H. Cornish (eds), *The Evolution of Language:*

Proceedings of the 10th International Conference (EVOLANG10), Vienna, Austria, 14–17 April 2014, 122–9. Singapore: World Scientific.

Isaac, A. M. C. (2019). 'The Allegory of Isomorphism'. *AVANT*, X (3): 1–23.

Ishai, A., L. G. Ungerleider and J. V. Haxby (2000). 'Distributed Neural Systems for the Generation of Visual Images'. *Neuron*, 28: 979–90.

Ison, M. J. and R. Q. Quiroga (2008). 'Selectivity and Invariance for Visual Object Perception'. *Frontiers in Bioscience*, 13: 4889–903.

Ittelson, W. H. (1996). 'Visual Perception of Markings'. *Psychonomic Bulletin & Review*, 3 (2): 171–87.

Izard, V., C. Sann, E. S. Spelke and A. Streri (2009). 'Newborn Infants Perceive Abstract Numbers'. *Proceedings of the National Academy of Sciences*, 106 (25): 10382–5.

Jakubowski, K. (2020). 'Musical Imagery'. In A. Abraham (ed.), *The Cambridge Handbook of the Imagination*, 187–206. Cambridge: Cambridge University Press.

Jäkel, F., M. Singh, F. A. Wichmann and M. H. Herzog (2016). 'An Overview of Quantitative Approaches in Gestalt Perception'. *Vision Research*, 126: 3–8.

Johnson, K. (2004). 'On the Systematicity of Language and Thought'. *Journal of Philosophy*, 101 (3): 111–39.

Johnson-Laird, P. N. (1998). 'Imagery, Visualization, and Thinking'. In J. Hochberg (ed.), *Perception and Cognition at the Century's End*, 441–67. San Diego, CA: Academic Press.

Kamermans, K. L., W. Pouw, F. W. Mast and F. Paas (2019). 'Reinterpretation in Visual Imagery Is Possible Without Visual Cues: A Validation of Previous Research'. *Psychological Research: An International Journal of Perception, Attention, Memory and Action*, 83 (6): 1237–50.

Kan, I. P., L. W. Barsalou, K. O. Solomon, J. K. Minor and S. L. Thompson-Schill (2003). 'Role of Mental Imagery in a Property Verification Task: fMRI Evidence for Perceptual Representations of Conceptual Knowledge'. *Cognitive Neuropsychology*, 20: 525–40.

Kant, I. (1998). *Critique of Pure Reason*. Trans. P. Guyer and A. Wood. Cambridge: Cambridge University Press.

Kaplan, D. (1968). 'Quantifying'. *Synthese*, 19 (1–2): 178–214.

Kaplan, D. (1989). 'Demonstratives'. In J. Almog, J. Perry, and H. Wettstein (eds), *Themes from Kaplan*, 481–563. New York: Oxford University Press.

Kasem, A. (1989). 'Can Berkeley Be Called an Imagist?' *Indian Philosophical Quarterly*, 6: 75–88.

Kaufmann, G. (1996). 'The Many Faces of Mental Imagery'. In C. Cornoldi, R. H. Logie, M. A. Brandimonte, G. Kaufmann and D. Reisberg (eds), *Stretching the Imagination: Representation and Transformation in Mental Imagery*, 77–118. New York: Oxford University Press.

Kennedy, J. M. (1974). *A Psychology of Picture Perception*. San Francisco; Washington; London: Jossey-Bass Publishers.

Keogh, R. and J. Pearson (2018). 'The Blind Mind: No Sensory Visual Imagery in Aphantasia'. *Cortex*, 105: 53–60.

Khalifa, K. (2017). *Understanding, Explanation, and Scientific Knowledge*. New York: Cambridge University Press.

Kiefer, M. and F. Pulvermüller (2012). 'Conceptual Representations in Mind and Brain: Theoretical Developments, Current Evidence and Future Directions'. *Cortex*, 48: 805–25.

Kind, A. (2018). 'How Imagination Gives Rise to Knowledge'. In F. Macpherson and F. Dorsch (eds), *Perceptual Memory and Perceptual Imagination*, 227–46. New York: Oxford University Press.

Kirsh, D. (2010). 'Thinking with External Representations'. *Artificial Intelligence and the Simulation of Behaviour*, 25: 441–54.

Kirsh, D. and P. Maglio (1994). 'On Distinguishing Epistemic from Pragmatic Action'. *Cognitive Science*, 18: 513–49.

Kitcher, P. and A. Varzi (2000). 'Some Pictures are Worth 2[aleph]0 Sentences'. *Philosophy*, 75 (3): 377–81.

Kjørup, S. (1974). 'George Inness and the Battle at Hastings, or Doing Things With Pictures'. *Monist*, 58 (2): 216–35.

Klatzky, R. L., S. J. Lederman and D. E. Matula (1991). 'Imagined Haptic Exploration in Judgements of Object Properties'. *Journal of Experimental Psychology: Learning, Memory, and Cognition*, 17: 314–22.

Knauff, M. (2013). *Space to Reason: A Spatial Theory of Human Thought*. Cambridge, MA; London: MIT Press.

Knauff, M. and P. N. Johnson-Laird (2002). 'Visual Imagery Can Impede Reasoning'. *Memory and Cognition*, 30: 363–71.

Kobayashi, M., M. Takeda, N. Hattori, M. Fukunaga, T. Sasabe, N. Inoue, Y. Nagai, T. Sawada, N. Sadato and Y. Watanabe (2004). 'Functional Imaging of Gustatory Perception and Imagery: "Top-Down" Processing of Gustatory Signals'. *NeuroImage*, 23: 1271–82.

Kosslyn S. M. (1980). *Image and Mind*. Cambridge, MA: Harvard University Press.

Kosslyn, S. M. (1994). *Image and Brain: The Resolution of the Imagery Debate*. Cambridge, MA: MIT Press.

Kosslyn, S. M., T. M. Ball and B. J. Reiser (1978). 'Visual Images Preserve Metric Spatial Information: Evidence from Studies of Image Scanning'. *Journal of Experimental Psychology: Human Perception and Performance*, 4: 47–60.

Kosslyn, S. M., W. L. Thompson and G. Ganis (2006). *The Case for Mental Imagery*. Oxford: Oxford University Press.

Kosso, P. (2007). 'Scientific Understanding'. *Foundations of Science*, 12: 173–88.

Kozak, P. (2020) 'The Diagram Problem'. In A.-V. Pietarinen, P. Chapman, L. Bosveld-de Smet, V. Giardino, J. Corter and S. Linker (eds), *Diagrammatic Representation and Inference*, 216–24. Cham: Springer.

Kozak, P. (2021). 'The Analog-Digital Distinction Fails to Explain the Perception-Thought Distinction: An Alternative Account of the Format of Mental Representation'. *Studia Semiotyczne*, 35 (1): 73–94.

Kozhevnikov, M., S. Kosslyn and J. Shephard (2005). 'Spatial vs. Object Visualizers: A New Characterization of Visual Cognitive Style'. *Memory and Cognition*, 33 (4): 710–26.

Kozhevnikov, M., O. Blazhenkova and M. Becker (2010). 'Trade-Off in Object Versus Spatial Visualization Abilities: Restriction in the Development of Visual-Processing Resources'. *Psychonomic Bulletin & Review*, 17 (1): 29–35.

Krantz, D., R. D. Luce, A. Tversky and P. Suppes (1971). *Foundations of Measurement Volume I: Additive and Polynomial Representations*. San Diego and London: Academic Press.

Kripke, S. A. (1980). *Naming and Necessity*. Cambridge, MA: Harvard University Press.

Kripke, S. A. (1982). *Wittgenstein on Rules and Private Language*. Oxford: Blackwell.

Kulpa, Z. (1987). 'Putting Order in the Impossible'. *Perception*, 16: 201–14.

Kulpa, Z. (2009). 'Main Problems of Diagrammatic Reasoning. Part I: The Generalization Problem'. *Foundations of Science*, 14 (1–2): 75–96.

Kulvicki, J. (2006a). *On Images: Their Structure and Content*. Oxford: Clarendon Press.

Kulvicki, J. (2006b). 'Pictorial Representation'. *Philosophy Compass*, 1 (6): 535–46.

Kulvicki, J. (2014). *Images*. New York: Routledge.

Kulvicki, J. (2020). *Modeling the Meanings of Pictures: Depiction and the Philosophy of Language*. Oxford: Oxford University Press.

Kvanvig, J. L. (2018). 'Knowledge, Understanding, and Reasons for Belief'. In D. Starr (ed.), *The Oxford Handbook of Reasons and Normativity*, 685–715. New York: Oxford University Press.

Lakoff, G. (1994). 'What is Metaphor?' In J. A. Barnden and K. J. Holyoak (eds), *Advances in Connectionist and Neural Computation Theory: Vol. 3, Analogy, Metaphor, and Reminding*, 203–57. Norwood, NJ: Ablex.

Lakoff, G. and M. Johnson (1980). *Metaphors We Live By*. Chicago, IL: University of Chicago Press.

Langacker, R. W. (1987). *Foundations of Cognitive Grammar*. Stanford, CA: Stanford University Press.

Langland-Hassan, P. (2015). 'Imaginative Attitudes'. *Philosophy and Phenomenological Research*, 40: 664–86.

Langland-Hassan, P. (2020). *Explaining Imagination*. Oxford: Oxford University Press.

Larkin, J. H. and H. A. Simon (1987). 'Why a Diagram is (Sometimes) Worth Ten Thousand Words'. *Cognitive Science*, 11: 65–99.

Latour, B. (1990). 'Visualisation and Cognition: Drawing Things Together'. In M. Lynch and S. Woolgar (eds), *Representation in Scientific Activity*, 19–68. Cambridge, MA: MIT Press.

Laurence, S. and E. Margolis (2012). 'The Scope of the Conceptual'. In E. Margolis, R. Samuels and S. Stich (eds), *The Oxford Handbook of Philosophy of Cognitive Science*, 291–317. New York: Oxford University Press.

Lehrer, K. (2000). 'Meaning, Exemplarization and Metarepresentation'. In D. Sperber (ed.), *Metarepresentations: A Multidisciplinary Perspective*, 299–310. Oxford: Oxford University Press.

Lemon, O., M. de Rijke and A. Shimojima (1999). 'Efficacy of Diagrammatic Reasoning'. *Journal of Logic, Language, and Information*, 8 (3): 265–71.

Lewis, D. (1970). 'General Semantics'. *Synthese*, 22: 18–67.

Lewis, D. (1971). 'Analog and Digital'. *Noûs*, 5: 321–8.

Lindsay, R. K. (1998). 'Using Diagrams to Understand Geometry'. *Computational Intelligence*, 14 (2): 222–56.

Lipton, P. (2009). 'Understanding Without Explanation'. In H. de Regt, S. Leonelli, and K. Eigner (eds), *Scientific Understanding: Philosophical Perspectives*, 43–63. Pittsburgh, PA: University of Pittsburgh.

List, C. J. (1981). 'Images, Propositions, and Natural Signs'. In E. Klinger (ed.), *Concepts, Results, and Applications*, 67–76. Boston, MA: Springer.

Liu, Z., D. C. Knill and D. Kersten (1995). 'Object Classification for Human & Ideal Observers'. *Vision Research*, 35: 549–68.

Locke, J. (1975). *An Essay Concerning Human Understanding*. Ed. P. H. Nidditch. Oxford: Oxford University Press.

Loewenstein, J. and D. Gentner (2005). 'Relational Language and the Development of Relational Mapping'. *Cognitive Psychology*, 50 (4): 315–53.

Lopes, D. M. (1996). *Understanding Pictures*. Oxford: Oxford University Press.

Lopes, D. M. (1997). 'Art Media and the Sense Modalities: Tactile Pictures'. *The Philosophical Quarterly*, 47: 425–40.

Lopes, D. M. (2003). 'Pictures and the Representational Mind'. *The Monist*, 86 (4): 632–52.

Lopes, D. M. (2005). *Sight and Sensibility: Evaluating Pictures*. Oxford: Oxford University Press.

Lopes, D. M. (2010). 'Picture This: Image-Based Demonstratives'. In C. Abell and K. Bantinaki (eds), *Philosophical Perspectives on Depiction*, 52–80. Oxford; New York: Oxford University Press.

Lowe, D. G. (2004). 'Distinctive Image Features From Scale-Invariant Keypoints'. *International Journal of Computer Vision*, 60 (2): 91–110.

Lowe, E. J. (1995). *Locke on Human Understanding*. London: Routledge.

Lowe, E. J. (1996). *Subjects of Experience*. Cambridge: Cambridge University Press.

Lupyan, G., D. H. Rakison and J. L. McClelland (2007). 'Language Is Not Just for Talking: Redundant Labels Facilitate Learning of Novel Categories'. *Psychological Science*, 18 (12): 1077–83.

Macbeth, D. (2010). 'Diagrammatic Reasoning in Euclid's Elements'. In B. Van Kerkhove, J. De Vuyst and J. P. Van Bendegem (eds), *Philosophical Perspectives on Mathematical Practice*, 235–67. London: College Publications.

Macbeth, D. (2012). 'Seeing How It Goes: Paper-and-Pencil Reasoning in Mathematical Practice'. *Philosophia Mathematica*, 20: 58–85.

Machery, E. (2007). 'Concept Empiricism: A Methodological Critique'. *Cognition*, 104: 19–46.

Machery, E. (2016). 'The Amodal Brain and the Offloading Hypothesis'. *Psychonomic Bulletin & Review*, 23: 1090–5.

MacKisack, M., S. Aldworth, F. Macpherson, J. Onians, C. Winlove and A. Zeman (2016). 'On Picturing a Candle: The Prehistory of Imagery Science'. *Frontiers in Psychology*, 7: 515.

Mahon, B. Z. (2015). 'What is Embodied About Cognition?' *Language, Cognition and Neuroscience*, 30: 420–9.

Maley, C. J. (2011). 'Analog and Digital, Continuous and Discrete'. *Philosophical Studies*, 155: 117–13.

Malinas, G. (1991). 'A Semantics for Pictures'. *Canadian Journal of Philosophy*, 21 (3): 275–98.

Mancosu, P. (2005). 'Visualization in Logic and Mathematics'. In P. Mancosu, K. Jørgensen and S. Pedersen (eds), *Visualization, Explanation and Reasoning Styles in Mathematics*, 13–30. Dordrecht: Springer.

Manders, K. (2008). 'The Euclidean Diagram'. In P. Mancosu (ed.), *The Philosophy of Mathematical Practice*, 80–133. Oxford: Oxford University Press.

Marcus, R. B. (1990). 'Some Revisionary Proposals About Belief and Believing'. *Philosophy and Phenomenological Research*, 50: 132–153.

Martin, A. (2007). 'The Representation of Object Concepts in the Brain'. *Annual Review of Psychology*, 58: 25–45.

Mast, F. W. and S. M. Kosslyn (2002). 'Visual Mental Images Can Be Ambiguous: Insights From Individual Differences in Spatial Transformation Abilities'. *Cognition*, 86: 57–70.

Matthen, M. (2005). *Seeing, Doing, and Knowing: A Philosophical Theory of Sense Perception*. Oxford: Clarendon Press.

Matthen, M. (2014). 'Image Content'. In B. Brogaard (ed.), *Does Perception Have Content?*, 265–90. New York: Oxford University Press.

Matthews, R. J. (2007). *The Measure of Mind: Propositional Attitudes and their Attribution*. Oxford: Oxford University Press.

Matthews, R. J. (2011). 'Measurement-Theoretic Accounts of Propositional Attitudes'. *Philosophy Compass*, 6 (11): 828–41.

Maynard, P. (2011). 'What Drawing Draws On: The Relevance of Current Vision Research'. *Rivista di estetica*, 47 (2): 9–29.

Maynard, P. (2017). 'Photo Mensura'. In N. Mößner and A. Nordmann (eds), *Reasoning in Measurement*, 41–56. London; New York: Routledge.

McCarthy, R. A. and E. K. Warrington (1986). 'Visual Associative Agnosia: A Clinico-Anatomical Study of a Single Case'. *Journal of Neurology Neurosurgery & Psychiatry*, 49 (11): 1233–40.

McDowell, J. (1996). *Mind and World*. Cambridge, MA: Harvard University Press.

McGinn, C. (1989). *Mental Content*. Oxford: Blackwell Publishers.

McGinn, C. (2006). *Mindsight: Image, Dream, Meaning*. Cambridge, MA: Harvard University Press.

McNeill, D. (1992). *Hand in Mind: What Gestures Reveal about Thought*. Chicago; London: Chicago University Press.

Mersch, D. (2003). 'Wort, Bild, Ton, Zahl: Medialitäten Medialen Darstellens'. In D. Mersch (ed.), *Die Medien der Künste*, 9–49. München: Wilhelm Fink Verlag.
Mersch, D. (2011). 'Aspects of Visual Epistemology: On the "Logic" of the Iconic'. In A. Benedek and K. Nyiri (eds), *Images in Language: Metaphors and Metamorphoses*, 169–94. Frankfurt a. M.: Peter Lang.
Merricks, T. (2009). 'Propositional Attitudes?'. *Proceedings of the Aristotelian Society*, 109: 207–32.
Meynell, L. (2018). 'Images and Imagination in Thought Experiments'. In M. Stuart, Y. Fehige and J. R. Brown (eds), *The Routledge Companion to Thought Experiments*, 498–511. New York: Routledge.
Meynell, L. (2020). 'Getting the Picture: Towards a New Account of Scientific Understanding'. In M. Ivanova and S. French (eds), *The Aesthetics of Science: Beauty, Imagination and Understanding*, 36–62. New York: Routledge.
Michaelian K. (2016). *Mental Time Travel: Episodic Memory and Our Knowledge of the Personal Past*. Cambridge, MA: MIT Press.
Mitchell, W. J. T. (1994). *Picture Theory: Essays on Verbal and Visual Representation*. Chicago; London: University of Chicago Press.
Molitor, S., S. Ballstaedt and H. Mandl (1989). 'Problems in Knowledge Acquisition From Text and Pictures'. In H. Mandl and J. R. Levin (eds), *Knowledge Acquisition from Text and Pictures*, 3–30. Amsterdam: North-Holland.
Montague, M. (2007). 'Against Propositionalism'. *Noûs*, 41 (3): 503–18.
Morrison, M. and M. S. Morgan (1999). 'Models as Mediating Instruments'. In M. S. Morgan and M. Morrison (eds), *Models as Mediators: Perspectives on Natural and Social Science*, 10–38. Cambridge: Cambridge University Press.
Mortensen, C. (2010). *Inconsistent Geometry*. London: College Publications.
Mortensen, C., S. Leishman, P. Quigley and T. Helke (2013). 'How Many Impossible Images Did Escher Produce?' *British Journal of Aesthetics*, 53 (4): 425–41.
Mößner, N. (2018). *Visual Representations in Science: Concept and Epistemology*. London; New York: Routledge.
Mullarkey, J. (2011). 'Temple Grandin's Animal Thoughts: On Non-Human Thinking in Pictures, Films, and Diagrams'. In A. Benedek and K. Nyiri (eds), *Images in Language: Metaphors and Metamorphoses*, 155–68. Frankfurt a. M.: Peter Lang.
Mumma, J. (2010). 'Proofs, Pictures, and Euclid'. *Synthese*, 175 (2): 255–87.
Murez, M. and F. Recanati (2016). 'Mental Files: An Introduction'. *Review of Philosophy and Psychology*, 7: 265–81.
Müller, A. (1997). *Die ikonische Differenz: Das Kunstwerk als Augenblick*. München: Wilhelm Fink Verlag.
Nail, T. (2019). *Theory of the Image*. New York: Oxford University Press.
Nanay, B. (2010). 'Perception and Imagination: Amodal Perception as Mental Imagery'. *Philosophical Studies*, 150: 239–54.
Nanay, B. (2014). *Aesthetics as Philosophy of Perception*. Oxford: Oxford University Press.

Nanay, B. (2015). 'Perceptual Content and the Content of Mental Imagery'. *Philosophical Studies*, 172: 1723–36.
Nanay, B. (2017). 'Pain and Mental Imagery'. *The Monist*, 100: 485–500.
Nanay, B. (2018). 'The Importance of Amodal Completion in Everyday Perception'. *i-Perception*, 9 (4): 1–16.
Nanay, B. (2021). 'Unconscious Mental Imagery'. *Philosophical Transactions of the Royal Society B*, 376 (1817): 20190689.
Narens, L. (1985). *Abstract Measurement Theory*. Cambridge, MA: MIT Press.
Neander, K. (1987). 'Pictorial Representation: A Matter of Resemblance'. *British Journal of Aesthetics*, 27 (3): 213–26.
Nelsen, R. B. (1993). *Proofs Without Words: Exercises in Visual Thinking*. Washington: The Mathematical Association of America.
Nersessian, N. (2008). *Creating Scientific Concepts*. Cambridge, MA: MIT Press.
Newell, A. and H. A. Simon (1972). *Human Problem Solving*. Englewood Cliffs, NJ: Prentice Hall.
Newall, M. (2011). *What is a Picture? Depiction, Realism, Abstraction*. London: Palgrave Macmillan.
Newton, N. (1982). 'Experience and Imagery'. *The Southern Journal of Philosophy*, 21: 475–87.
Novitz, D. (1977). *Pictures and Their Use in communication: A Philosophical Essay*. The Hague: Martinus Nijhoff.
Novitz, D. (1988). 'Review of Deeper into Pictures: An Essay on Pictorial Representation'. *British Journal of Aesthetics*, 28 (1): 87–9.
Nunberg, G. (1993). 'Indexicality and Deixis'. *Linguistics and Philosophy*, 16: 1–43.
Nyíri, J. C. (2001). 'The Picture Theory of Reason'. In B. Brogaard and B. Smith (eds), *Rationality and Irrationality* (Schriftenreihe-Wittgenstein Gesellschaft, Vol. 29), 242–66. Vienna: Öbv&hpt.
Nyíri, K. (2013). 'Wittgenstein's Philosophy of Pictures'. In A. Pichler and S. Säätelä (eds), *Wittgenstein: The Philosopher and his Works*, 322–53. Berlin; Boston: De Gruyter.
Otis, L. (2015). *Rethinking Thought: Inside the Minds of Creative Scientists and Artists*. Oxford; New York: Oxford University Press.
Overgaard, S. (2022). 'Seeing-as, Seeing-o, and Seeing-that'. *Philosophical Studies*, 179 (11): 1–20.
Pagin, P. and D. Westerståhl (2010). 'Compositionality I: Definitions and Variants'. *Philosophy Compass*, 5: 250–64.
Paivio, A. (1963). 'Learning of Adjective-Noun Paired Associates as a Function of Adjective-Noun Word Order and Noun Abstractness'. *Canadian Journal of Psychology*, 17: 370–79.
Paivio, A. and R. Harshman (1983). 'Factor Analysis of a Questionnaire on Imagery and Verbal Habits and Skills'. *Canadian Journal of Psychology*, 37: 461–83.
Peacocke, C. (1986). *Thoughts: An Essay on Content*. Oxford: Basil Blackwell.

Peacocke, C. (1987). 'Depiction'. *Philosophical Review*, 96 (3): 383–410.
Peacocke, C. (1992). *A Study of Concepts*. Cambridge, MA: MIT Press.
Peacocke, C. (2019). *The Primacy of Metaphysics*. Oxford: Oxford University Press.
Pearson, J. (2020). 'Reply to: Assessing the Causal Role of Early Visual Areas in Visual Mental Imagery'. *Nature Reviews Neuroscience*, 21: 517.
Pearson, J. and S. M. Kosslyn (2015). 'The Heterogeneity of Mental Representation: Ending the Mental Imagery Debate'. *Proceedings of the National Academy of Sciences PNAS (PNAS)*, 11: 10089–92.
Pearson, J., T. Naselaris, E. A. Holmes and S. M. Kosslyn (2015). 'Mental Imagery: Functional Mechanisms and Clinical Applications'. *Trends in Cognitive Sciences*, 19 (10): 590–602.
Pecher, D., R. Zeelenberg and L. W. Barsalou (2003). 'Verifying Different-Modality Properties for Concepts Produces Switching Costs'. *Psychological Science*, 14: 119–24.
Pecher, D., R. Zeelenberg and L. W. Barsalou (2004). 'Sensorimotor Simulations Underlie Conceptual Representations: Modality-Specific Effect of Prior Act'. *Psychonomic Bulletin & Review*, 11: 164–7.
Peirce, C. S. (1931–1935). *Collected Papers of Charles Sanders Peirce*, Vol. 1–6. Ed. C. Hartshorne and P. Weiss. Cambridge, MA: Harvard University Press (Cited as *CP* n. m where *n* and *m* indicate volume and paragraph numbers).
Peirce, C. S. (1976). *The New Elements of Mathematics*, Vol. 1–4. Ed. C. Eisele. Hague: Mouton Publishers; Atlantic Highlands, NJ: Humanities Press (Cited as *NEM* n. m where *n* and *m* indicate volume and page numbers).
Perini, L. (2005). 'The Truth in Pictures'. *Philosophy of Science*, 72: 262–85.
Perini, L. (2012). 'Truth-Bearers or Truth-Makers?' *Spontaneous Generations: A Journal for the History and Philosophy of Science*, 6 (1): 142–7.
Pero, F. and M. Suárez (2016). 'Varieties of Misrepresentation and Homomorphism'. *European Journal for Philosophy of Science*, 6 (1): 71–90.
Perry, J. (1979). 'The Problem of the Essential Indexical'. *Noûs*, 13 (1): 3–21.
Peterson, M. A., J. F. Kihlstrom, P. M. Rose and M. L. Glisky (1992). 'Mental Images Can Be Ambiguous: Reconstruals and Reference Frame Reversals'. *Memory and Cognition*, 20: 107–23.
Pietarinen, A.-V. (2006). *Signs of Logic: Peircean Themes on the Philosophy of Language, Games, and Communication*. Dordrecht: Springer.
Pietarinen, A.-V. (2014). 'A Road to the Philosophy of Iconic Communication'. In A. Benedek and K. Nyíri (eds), *The Power of the Image: Emotion, Expression, Explanation*, 47–60. Frankfurt a. M.: Peter Lang.
Pietarinen, A.-V. (2016). 'Is There a General Diagram Concept?' In S. Krämer and C. Ljungberg (eds), *Thinking with Diagrams: The Semiotic Basis of Human Cognition*, 121–37. Boston; Berlin: Walter de Gruyter.
Podro, M. (1998). *Depiction*. London: Yale University Press.
Price, H. H. (1953). *Thinking and Experience*. London: Hutchinson's University Library.

Priest, G. (1997). 'Sylvan's Box: A Short Story and Ten Morals'. *Notre Dame Journal of Formal Logic*, 38 (4): 573–82.
Priest, G. (1999). 'Perceiving Contradictions'. *Australasian Journal of Philosophy*, 77 (4): 439–46.
Prinz, J. J. (2002). *Furnishing the Mind: Concepts and Their Perceptual Basis*. Cambridge, MA: MIT Press.
Proclus (1992). *A Commentary on the First Book of Euclid's Elements*. Trans. G. Morrow. Princeton, NJ: Princeton University Press.
Putnam, H. (1981). *Reason, Truth, and History*. Cambridge: Cambridge University Press.
Pylyshyn, Z. W. (1973). 'What the Mind's Eye Tells the Mind's Brain: A Critique of Mental Imagery'. *Psychological Bulletin*, 80: 1–25.
Pylyshyn, Z. W. (2002). 'Mental Imagery: In Search of a Theory'. *Behavioral and Brain Sciences*, 25: 157–82.
Pylyshyn, Z. W. (2003a). *Seeing and Visualizing: It's Not What You Think*. Cambridge, MA: MIT Press.
Pylyshyn, Z. W. (2003b). 'Return of the Mental Image: Are There Really Pictures in the Brain?' *TRENDS in Cognitive Sciences*, 7 (3): 113–18.
Pylyshyn, Z. W. (2007). *Things and Places: How the Mind Connects with the World*. Cambridge, MA: MIT Press.
Quilty-Dunn, J. (2016). 'Iconicity and the Format of Perception'. *Journal of Consciousness Studies*, 23 (3–4): 255–63.
Quilty-Dunn, J. (2020). 'Perceptual Pluralism'. *Noûs*, 54 (4): 807–38.
Quilty-Dunn, J. and E. Mandelbaum (2020). 'Non-Inferential Transitions: Imagery and Association'. In T. Chan and A. Nes (eds), *Inference and Consciousness*, 151–71. New York: Routledge.
Quine, W. V. O. (1956). 'Quantifiers and Propositional Attitudes'. *Journal of Philosophy*, 5: 177–87.
Quine, W. V. O. (1960). *Word and Object*. Cambridge, MA: MIT Press.
Quine, W. V. O. (1968). 'Ontological Relativity'. *Journal of Philosophy*, 65: 185–212.
Quine, W. V. O. (1970). 'Methodological Reflections on Current Linguistic Theory'. *Synthese*, 21 (3–4): 386–98.
Raftopoulos, A. (2009). *Cognition and Perception: How do Psychology and Neural Science Inform Philosophy?* Cambridge, MA: MIT Press.
Raftopoulos, A. and J. Zeimbekis (2015). 'The Cognitive Penetrability of Perception – An Overview'. In J. Zembeiskis and A. Raftopolous (eds), *The Cognitive Penetrability of Perception: New Philosophical Perspectives*, 1–56. Oxford: Oxford University Press.
Ragni, M. and M. Knauff (2013). 'A Theory and a Computational Model of Spatial Reasoning with Preferred Mental Models'. *Psychological Review*, 120 (3): 561–88.
Rayo, A. (2013). *The Construction of Logical Space*. Oxford: Oxford University Press.
Recanati, F. (2012). *Mental Files*. New York: Oxford University Press.

Reisberg, D. (1996). 'The Nonambiguity of Mental Images'. In C. Cornoldi, R. H. Logie, M. A. Brandimonte, G. Kaufmann and D. Reisberg (eds), *Stretching the Imagination: Representation and Transformation in Mental Imagery*, 119–72. New York: Oxford University Press.

Reisberg, D. and D. Chambers (1991). 'Neither Pictures Nor Propositions: What We Can Learn From a Mental Image?' *Canadian Journal of Psychology*, 45: 336–52.

Reisberg, D. and F. Heuer (2005). 'Visuospatial Images'. In P. Shah and A. Miyake (eds), *The Cambridge Handbook of Visuospatial Thinking*, 35–80. New York: Cambridge University Press.

Reisberg, D., D. G. Pearson and S. M. Kosslyn (2003). 'Intuitions and Introspections About Imagery: The Role of Imagery Experience in Shaping an Investigator's Theoretical Views'. *Applied Cognitive Psychology*, 17: 147–60.

Rescorla, M. (2009a). 'Cognitive Maps and the Language of Thought'. *British Journal for the Philosophy of Science*, 60: 377–407.

Rescorla, M. (2009b). 'Predication and Cartographic Representation'. *Synthese*, 169 (1): 175–200.

Rey, G. (1995). 'A Not "Merely Empirical" Argument for the Language of Thought'. *Philosophical Perspectives*, 9: 201–22.

Rheingold, H. R. (1991). *Virtual Reality*. London: Secker and Warburg.

Richardson, J. T. E. (1987). 'The Role of Mental Imagery in Models of Transitive Inference'. *British Journal of Psychology*, 78: 189–203.

Richler, J. J., J. B. Wilmer and I. Gauthier (2017). 'General Object Recognition is Specific: Evidence from Novel and Familiar Objects'. *Cognition*, 166: 42–55.

Richtmeyer, U. (2019). *Wittgensteins Bilddenken*. Leiden, Niederlande: Wilhelm Fink.

Rocke, A. J. (2010). *Image and Reality: Kekulé, Kopp, and the Scientific Imagination*. Chicago, IL: University of Chicago Press.

Rollins, M. (1989). *Mental Imagery: On the Limits of Cognitive Science*. New Haven, CT: Yale University Press.

Roser, A. (1996). 'Gibt es Autonome Bilder? Bemerkungen zum Grafischen Werk Otto Neuraths und Ludwig Wittgensteins'. *Grazer Philosophische Studien*, 52 (1): 9–43.

Rozemond, M. (1993). 'Evans on *de re* Thought'. *Philosophia*, 22 (3–4): 275–98.

Russell, B. (1919). 'On Propositions: What They Are and How They Mean'. *Aristotelian Society Supplementary Volume*, 2: 1–43.

Russell, B. (1921). *The Analysis of Mind*. London: Allen & Unwin.

Russell, B. (1997). *The Problems of Philosophy*. Oxford: Oxford University Press.

Ryle, G. (1949). *The Concept of Mind*. London: Hutchinson.

Ryle, G. (1979). *On Thinking*. Oxford: Basil Blackwell.

Ryle, G. (2009). *Collected Essays 1929-1968: Collected Papers*, Vol. 2. Ed. J. Tanney. London; New York: Routledge.

Sainsbury, R. M. (2005). *Reference Without Referents*. Oxford: Oxford University Press.

Salis, F. and R. Frigg (2020). *Capturing the Scientific Imagination*. In A. Levy and P. Godfrey-Smith (eds), *The Scientific Imagination: Philosophical and Psychological Perspectives*, 17–50. New York: Oxford University Press.
Salje, L. (2019). 'Talking Our Way to Systematicity'. *Philosophical Studies*, 176: 2563–88.
Sartre, J.-P. (1962). *Imagination: A Psychological Critique*. Ann Arbor, MI: University of Michigan Press.
Sartwell, C. (1991). 'Natural Generativity and Imitation'. *British Journal of Aesthetics*, 31 (1): 58–67.
Scaife, M. and Y. Rogers (1996). 'External Cognition: How Do Graphical Representations Work?' *International Journal of Human-Computer Studies*, 45: 185–213.
Scarry, E. (1999). *Dreaming by the Book*. Princeton, NJ: Princeton University Press.
Schier, F. (1986). *Deeper Into Pictures*. Cambridge: Cambridge University Press.
Schooler, J. W., M. Fallshore and S. M. Fiore (1995). 'Epilogue: Putting Insight Into Perspective'. In R. J. Sternberg and J. E. Davidson (eds), *The Nature of Insight*, 559–88. Cambridge, MA; London: MIT Press.
Schöttler, T. (2017). 'Pictorial Evidence: On the Rightness of Pictures'. In N. Mößner and A. Nordmann (eds), *Reasoning in Measurement*, 111–29. London; New York: Routledge.
Sebeok, T. A. (1979). *The Sign & Its Masters*. Austin: University of Texas Press.
Sellars, W. (1949). 'Language, Rules and Behavior'. In S. Hook (ed.), *John Dewey: Philosopher of Science and Freedom*, 289–315. New York: Dial Press.
Sellars, W. (1963). *Science, Perception and Reality*. Atascadero: Ridgeview.
Sellars, W. (1969). 'Language as Thought and as Communication'. *Philosophy and Phenomenological Research*, 29 (4): 506–27.
Sellars, W. (2007). *In the Space of Reasons: Selected Essays of Wilfrid Sellars*. Ed. K. Scharp, R. B. Brandom. Cambridge, MA: Harvard University Press.
Shabel, L. (2003). *Mathematics in Kant's Critical Philosophy: Reflections on Mathematical Practice*. New York: Routledge.
Shah, P. and A. Miyake (1996). 'The Separability of Working Memory Resources for Spatial Thinking and Language Processing'. *Journal of Experimental Psychology: General*, 125: 4–27.
Shaver, P., L. Pierson and S. Lang (1975). 'Converging Evidence for the Functional Significance of Imagery in Problem Solving'. *Cognition*, 3: 359–75.
Shepard, R. N. (1978). 'Externalization of Mental Images and the Act of Creation'. In B. S. Randhawa and W. E. Coffman (eds), *Visual Learning, Thinking, and Communication*, 133–89. New York: Academic Press.
Shepard, R. N. (1990). 'On Understanding Mental Images'. In H. Barlow, C. Blakemore and M. Westo-Smith (eds), *Images and Understanding: Thoughts about Images: Ideas about Understanding*, 365–70. Cambridge: Cambridge University Press.
Shepard, R. N. and L. A. Cooper (1982). *Mental Images and Their Transformations*. Cambridge, MA: MIT Press.

Shepard, R. N. and J. Metzler (1971). 'Mental Rotation of Three-Dimensional Objects'. *Science*, 171: 701–3.
Sheredos, B., D. Burnston, A. Abrahamsen and W. Bechtel (2013). 'Why Do Biologists Use So Many Diagrams?' *Philosophy of Science*, 80 (5): 931–44.
Sheredos, B. and W. Bechtel (2020). 'Imagining Mechanisms with Diagrams'. In A. Levy and P. Godfrey-Smith (eds), *The Scientific Imagination: Philosophical and Psychological Perspectives*, 178–219. New York: Oxford University Press.
Shimojima, A. (1996). 'Operational Constraints in Diagrammatic Reasoning'. In J. Barwise and G. Allwein (eds), *Logical Reasoning with Diagrams*, 27–48. Oxford: Oxford University Press.
Shimojima, A. (2001). 'The Graphic-Linguistic Distinction: Exploring Alternatives'. In: A. Blackwell (ed.), *Thinking with Diagrams*, 5–27. Dordrecht: Kluwer Academic Publishers.
Shimojima, A. (2015). *Semantic Properties of Diagrams and Their Cognitive Potentials*. Stanford, CA: CSLI Publications.
Shin, S.-J. (1994). *The Logical Status of Diagrams*. New York: Cambridge University Press.
Shin, S.-J. (2002). *The Iconic Logic of Peirce's Graphs*. Cambridge, MA: MIT Press.
Shin, S.-J.(2015). 'The Mystery of Deduction and Diagrammatic Aspects of Representation'. *Review of Philosophy and Psychology*, 6 (1): 49–67.
Shin, S.-J. (2016). 'The Role of Diagrams in Abductive Reasoning'. In S. Krämer and C. Ljungberg (eds), *Thinking with Diagrams: The Semiotic Basis of Human Cognition*, 57–76. Boston; Berlin: Walter de Gruyter.
Short, T. S. (2007). *Peirce's Theory of Signs*. Cambridge: Cambridge University Press.
Siewert, C. (1998). *The Significance of Consciousness*. Princeton, NJ: Princeton University Press.
Skinner, B. F. (1957). *Verbal Behavior*. New York: Appleton.
Slezak, P. (1995). 'The "Philosophical" Case Against Visual Imagery'. In P. Slezak, T. Caelli and R. Clark (eds), *Perspectives on Cognitive Science: Theories, Experiments and Foundations*, 237–71. Norwood, NJ: Ablex Publishing.
Slezak, P. (2002a). 'The Imagery Debate: Déjà vu All Over Again?' *Behavioral and Brain Sciences*, 25 (2): 209–10.
Slezak, P. (2002b). 'Thinking About Thinking: Language, Thought and Introspection'. *Language & Communication*, 22: 353–73.
Sloman, A. (1978). *The Computer Revolution in Philosophy: Philosophy, Science, and Models of Mind*. New Jersey: Humanities Press.
Sloman, A. (2002). 'Diagrams in the Mind?' In M. Anderson, B. Meyer and P. Olivier (eds), *Diagrammatic Representation and Reasoning*, 7–28. London: Springer-Verlag.
Sober, E. (1976). 'Mental Representations'. *Synthese*, 33: 101–48.
Soles, D. (1999). 'Is Locke an Imagist?' *The Locke Newsletter*, 30: 17–66.
Solomon, K. O., D. L. Medin and E. B. Lynch (1999). 'Concepts Do More Than Categorize'. *Trends in Cognitive Sciences*, 3: 99–105.

Solt, K. (1989). 'Pictures and Truth'. *British Journal of Aesthetics*, 29 (2): 154–9.
Sorensen, R. (2002). 'The Art of the Impossible'. In T. S. Gendler and J. Hawthorne (eds), *Conceivability and Possibility*, 337–68. Oxford: Oxford University Press.
Spelke, E. S. and K. D. Kinzler (2007). 'Core Knowledge'. *Developmental Science*, 10: 89–96.
Squires, R. (1969). 'Depicting'. *Philosophy*, 44: 193–204.
Stanley, J. and T. Williamson (2001). 'Knowing How'. *The Journal of Philosophy*, 98 (8): 411–44.
Stegmüller, W. (1969). *Wissenschaftliche Erklärung und Begründung*. Berlin: Springer Verlag.
Stenning, K. (2002). *Seeing Reason: Image and Language in Learning to Think*. Oxford: Oxford University Press.
Stenning, K. and O. Lemon (2001). 'Aligning Logical and Psychological Perspectives on Diagrammatic Reasoning'. *Artificial Intelligence Review*, 15 (1–2): 29–62.
Stenning, K. and J. Oberlander (1995). 'A Cognitive Theory of Graphical and Linguistic Reasoning: Logic and Implementation'. *Cognitive Science*, 19: 97–140.
Stenning, K. and P. Yule (1997). 'Image and Language in Human Reasoning: A Syllogistic Illustration'. *Cognitive Psychology*, 34 (2): 109–59.
Sternberg, R. J. (1980). 'Representation and Process in Linear Syllogistic Reasoning'. *Journal of Experimental Psychology: General*, 109 (2): 119–59.
Stich, S. P. (1978). 'Beliefs and Subdoxastic States'. *Philosophy of Science*, 45 (4): 499–518.
Stjernfelt, F. (2000). 'Diagrams as Centerpiece of a Peircean Epistemology'. *Transactions of the Charles S. Peirce Society*, 36 (3): 357–84.
Stjernfelt, F. (2007). *Diagrammatology: An Investigation on the Borderlines of Phenomenology, Ontology and Semiotics*. Dordrecht: Springer.
Strawson, P. F. (1963). *Individuals: An Essay in Descriptive Metaphysics*. New York: Anchor Books.
Suárez, M. (2003). 'Scientific Representation: Against Similarity and Isomorphism'. *International Studies in the Philosophy of Science*, 17 (3): 225–44.
Swoyer, C. (1987). 'The Metaphysics of Measurement'. In J. Forge (ed.), *Measurement, Realism and Objectivity: Essays on Measurement in the Social and Physical Sciences*, 235–90. Dordrecht: Springer.
Swoyer, C. (1991). 'Structural Representation and Surrogative Reasoning'. *Synthese*, 87 (3): 449–508.
Tal, E. (2013). 'Old and New Problems in Philosophy of Measurement'. *Philosophy Compass*, 8 (12): 1159–73.
Tal, E. (2017). 'A Model-Based Epistemology of Measurement'. In N. Mößner and A. Nordmann (eds), *Reasoning in Measurement*, 233–53. London; New York: Routledge.
Tarski, A. (1956). *Logic, Semantics, Metamathematics: Papers from 1923 to 1938*. Oxford: Oxford University Press.
Teller, P. (2013). 'The Concept of Measurement-Precision'. *Synthese*, 190: 189–202.

Teller, P. (2018). 'Measurement Accuracy Realism'. In I. Peschard and B. C. van Fraassen (eds), *The Experimental Side of Modeling*, 273–98. Minneapolis, MN: University of Minnesota Press.

Terrone, E. (2021a). 'Seeing-in and Singling Out: How to Reconcile Pictures with Singular Thought'. *Pacific Philosophical Quarterly*, 102 (3): 378–92.

Terrone, E. (2021b). 'The Standard of Correctness and the Ontology of Depiction'. *American Philosophical Quarterly*, 58 (4): 399–412.

Teufel, C. and B. Nanay (2017). 'How to (And How Not To) Think About Top-Down Influences on Perception'. *Consciousness and Cognition*, 47: 17–25.

Thagard, P. (2005). *Mind: Introduction to Cognitive Science*. Cambridge, MA; London: MIT Press.

Thagard, P. (2012). *The Cognitive Science of Science: Explanation, Discovery, and Conceptual Change*. Cambridge, MA; London: MIT Press.

Thielen, J., S. E. Bosch, T. M. van Leeuwen, M. van Gerven and R. van Lier (2019). 'Neuroimaging Findings on Amodal Completion: A Review'. *i-Perception*, 10 (2): 1–25.

Thomas, N. J. T. (1997). 'A Stimulus to the Imagination'. *PSYCHE: An Interdisciplinary Journal of Research On Consciousness*, 3 (4): 1–14.

Thomas, N. J. T. (2003). 'Mental Imagery, Philosophical Issues About'. In L. Nadel (ed.), *Encyclopedia of Cognitive Science*, Vol. 2, 147–1153. London: Nature Publishing/Macmillan.

Thomas, N. J. T. (2009). 'Visual Imagery and Consciousness'. In W. P. Banks (ed.), *Encyclopedia of Consciousness*, Vol. 2, 445–57. Oxford: Academic Press.

Tichy, P. (1986). 'Constructions'. *Philosophy of Science*, 53: 514–34.

Tichy, P. (2012). *The Foundations of Frege's Logic*. Berlin; Boston: De Gruyter.

Troscianko, E. T. (2013). 'Reading Imaginatively: The Imagination in Cognitive Science and Cognitive Literary Studies'. *Journal of Literary Semantics*, 42 (2): 181–98.

Tufte, E. R. (1997). *Visual Explanations: Images and Quantities, Evidence and Narrative*. Cheshire, CN: Graphics Press.

Tversky, B. (2004). 'Semantics, Syntax, and Pragmatics of Graphics'. In K. Holmqvist and Y. Ericsson (eds), *Language and Visualization*, 141–58. Lund: Lund University Press.

Tversky, B. (2011). 'Visualizing Thought'. *Topics in Cognitive Science*, 3: 499–535.

Tversky, B. (2015). 'The Cognitive Design of Tools of Thought'. *Review of Philosophy and Psychology*, 6: 99–116.

Twardowski, K. (1965). 'Wyobrażenia i pojęcia'. In K. Twardowski (ed.), *Wybrane Pisma Filozoficzne*, 114–97. Warszawa: PWN.

Twardowski, K. (1999). *On Actions, Products and Other Topics in Philosophy*. Ed. J. Brandl and J. Wolenski. Amsterdam: Rodopi.

Tye, M. (2005). 'Non-Conceptual Content, Richness, and Fineness of Grain'. In T. Gendler and J. Hawthorne (eds), *Perceptual Experience*, 504–26. Oxford: Oxford University Press.

Tye, M. (2012). 'Knowing What It Is Like'. In J. Bengson and M. A. Moffett (eds), *Knowing How: Essays on Knowledge, Mind, and Action*, 300–13. Oxford: Oxford University Press.

Vallée-Tourangeau, F., S. V. Steffensen, G. Vallée-Tourangeau and M. Sirota (2016). 'Insight with Hands and Things'. *Acta Psychologica*, 170: 195–205.

Van Fraassen, B. C. (2008). *Scientific Representation: Paradoxes of Perspective*. Oxford: Oxford University Press.

Vernazzani, A. (2021). 'Do We See Facts?' *Mind & Language*, 37 (4): 674–93.

Viera, G. and Nanay, B. (2020). 'Temporal Mental Imagery'. In A. Abraham (ed.), *The Cambridge Handbook of the Imagination*, 227–40. Cambridge: Cambridge University Press.

Voltolini, A. (2015). *A Syncretistic Theory of Depiction*. New York: Palgrave Macmillan.

Von Neumann, J. (1958). *The Computer and the Brain*. New Haven, CT: Yale University Press.

Walton, K. L. (1973). 'Pictures and Make-Believe'. *The Philosophical Review*, 82 (3): 283–319.

Walton, K. L. (1984). 'Transparent Pictures: On the Nature of Photographic Realism'. *Critical Inquiry*, 11 (2): 246–77.

Walton, K. L. (1990). *Mimesis as Make-Believe*. Cambridge, MA: Harvard University Press.

Ware, C. (2000). *Information Visualization: Perception for Design*. San Francisco, CA: Morgan Kaufmann.

Warrington, E. K. and M. James (1988). 'Visual Apperceptive Agnosia: A Clinico-Anatomical Study of Three Cases'. *Cortex: A Journal Devoted to the Study of the Nervous System and Behavior*, 24 (1): 13–32.

Watson, J. B. (1913). 'Psychology as the Behaviorist Views It'. *Psychological Review*, 20: 158–77.

Westerhoff, J. (2005). 'Logical Relations between Pictures'. *Journal of Philosophy*, 102 (12): 603–23.

White, A. R. (1990). *The Language of Imagination*. Oxford: Blackwell.

Wiesing, L. (2009). *Artificial Presence: Philosophical Studies in Image Theory*. Trans. N. F. Schott. Stanford, CA: Stanford University Press.

Willats, J. (1997). *Art and Representation: New Principles in the Analysis of Pictures*. Princeton, NJ: Princeton University Press.

Williamson, T. (2002). *Knowledge and Its Limits*. Oxford: Oxford University Press.

Williamson, T. (2016). 'Knowing by Imagining'. In A. Kind and P. Kung (eds), *Knowledge Through Imagination*, 113–23. New York: Oxford University Press.

Winn, W. (1993). 'An Account of How Readers Search for Information in Diagrams'. *Contemporary Educational Psychology*, 18: 162–85.

Wittgenstein, L. (1953). *Philosophical Investigations*. Ed. G. E. M. Anscombe and R. Rhees, trans. G. E. M. Anscombe, 2nd edn., 1958. Oxford: Blackwell.

Wittgenstein, L. (1958). *The Blue and Brown Books [1933–1935]*. Oxford: Blackwell.

Wollheim, R. (1980). *Art and its Objects: Second Edition with Six Supplementary Essays.* Cambridge: Cambridge University Press.
Wollheim, R. (1984). *The Thread of Life.* Cambridge: Cambridge University Press.
Wollheim, R. (1987). *Painting as an Art.* Princeton, NJ: Princeton University Press.
Wollheim, R. (1993). *The Mind and its Depths.* Cambridge, MA; London: Harvard University Press.
Wollheim, R. (1998). 'On Pictorial Representation'. *The Journal of Aesthetics and Art Criticism*, 56 (3): 217–26.
Wollheim, R. (2003). 'What Makes Representational Painting Truly Visual?' *Aristotelian Society Supplementary Volume*, 77 (1): 131–47.
Yolton, J. W. (1996). *Perception and Reality: A History from Descartes to Kant.* Ithaca, NY: Cornell University Press.
Yoo, S.-S., D. K. Freeman, J. J. III McCarthy and F. A. Jolesz (2003). 'Neural Substrates of Tactile Imagery: A Functional MRI Study'. *Neuroreport*, 14 (4): 581–5.
Zacks, J. M., E. Levy, B. Tversky and D. J. Schiano (2002). 'Graphs in Print'. In M. Anderson, B. Meyer and P. Olivier (eds), *Diagrammatic Representation and Reasoning*, 187–206. London: Springer.
Zagzebski, L. (2019). 'Toward a Theory of Understanding'. In S. R. Grimm (ed.), *Varieties of Understanding: New Perspectives from Philosophy, Psychology, and Theology*, 123–36. New York: Oxford University Press.
Zatorre, R. J., A. R. Halpern and M. Bouffard (2010). 'Mental Reversal of Imagined Melodies: A Role for the Posterior Parietal Cortex'. *Journal of Cognitive Neuroscience*, 22 (4): 775–89.
Zeimbekis, J. (2010). 'Pictures and Singular Thought'. *The Journal of Aesthetics and Art Criticism*, 68 (1): 11–21.
Zeimbekis, J. (2015). 'Seeing, Visualizing, and Believing: Pictures and Cognitive Penetration'. In J. Zeimbekis and A. Raftopoulos (eds), *The Cognitive Penetrability of Perception: New Philosophical Perspectives*, 298–327. Oxford: Oxford University Press.
Zeman, A. Z. J., S. Della Sala, L. A. Torrens, V.-E. Gountouna, D. J. McGonigle and R. H. Logie (2010). 'Loss of Imagery Phenomenology with Intact Visuo-Spatial Task Performance: A Case of "Blind Imagination"'. *Neuropsychologia*, 48: 145–55.
Zeman, A., M. Dewar and S. Della Sala (2015). 'Lives without Imagery – Congenital Aphantasia'. *Cortex*, 73: 378–80.
Zeman, A., F. Milton, S. Della Sala, M. Dewar, T. Frayling, J. Gaddum, A. Hattersley, B. Heuerman-Williamson, K. Jones, M. MacKisack and C. Winlove (2020). 'Phantasia – The Psychological Significance of Lifelong Visual Imagery Vividness Extremes'. *Cortex*, 130: 426–40.
Zhang, J. (1997). 'The Nature of External Representations in Problem Solving'. *Cognitive Science*, 21: 179–217.
Zhao, F., W. Schnotz, I. Wagner and R. Gaschler (2020). 'Texts and Pictures Serve Different Functions in Conjoint Mental Model Construction and Adaptation'. *Memory and Cognition*, 48 (1): 69–82.

Index

Aasen, S. 105
Abel, G. 19
Abell, C. 32, 96, 126, 185 n.4
Abid, G. 179
Abrahamsen, A. 17
algebra 77, 190 n.4
algorithm 77, 79, 85–6
Ambrosio, C. 9
Amit E. 8
amodal completion 178–9, 195 n.8
Anderson, J. R. 195 n.5
Antonietti, A. 191 n.10
Aphantasia 177–8, 185 n.3
Armstrong, D. 107, 131, 186 n.2, 191 n.2
Arnheim, R. 187 n.1
associationism (associationist theory of thought) 46–50, 187 n.8
Ayers, M. 188 n.2

Baddeley, A. D. 56
Baillargeon, R. 189 n.3
Ball, D. N. 137
Ballung concepts 138
Bantinaki, K. 32, 126
Barsalou, L. W. 19, 57–8, 60
Bartolomeo, P. 195 n.6
Barwise, J. 7, 22
Bauer, M. I. 21
Bayne, T. 27
Beauchamp, M. S. 189 n.4
Bechtel, W. 17, 19, 185 n.4
Beck, J. 169
Bedny, M. 189 n.6
behaviourism 18, 47, 50–1
Beilock, S. L. 22
Belting, H. 3
Bennett, M. 18, 64–5, 69
Bensafi, M. 10
Ben-Yami, H. 13
Berman, D. 7
Bermúdez, J. L. 12–13, 110
Bernecker, S. 192 n.4

Bildobjekt 131, 193 n.2
binding problem 42
bistable figures 110, 174, 178
Black, M. 191 n.11
black hole picture 16, 75, 101–3, 105–10, 119–20, 122, 127–8, 132–3, 151–2, 154–6, 165, 182
Blackwell, S. E. 10
Block, N. 19, 175
Blumson, B. 32, 49, 96, 144
Bocanegra, B. R. 6
Boehm, G. 3
Bordwell, D. 22
Botterill, G. 17
Boumans, M. 138
Bourne, L. E. 186 n.1
Bradburn, N. M. 137
Braddon-Mitchell, D. 45
Braine, M. D. S. 12
Briscoe, R. 11
Brown, J. R. 7
Buccella, A. 114
Budd, M. 11, 105, 144
Bueno, O. 95–6
Burge, T. 9, 19, 41, 58, 103, 164–5, 194 n.3
Burke, L. 108
Byrne, R. M. J. 6

calibration 141–2
Camp, E. 25, 45, 49, 165, 171, 187 n.6
Candlish, S. 19
canonical decomposition 43–5, 164–5, 194 n.3
Cantlon, J. F. 189 n.6
Carey, S. 19, 61, 189 n.3
Carruthers, P. 17
Cartwright, N. 137, 150
Carvalho, J. M. 5
Casati, R. 21, 106, 164
Cavedon-Taylor, D. 175
Chambers, D. 174
Chang, H. 135, 137, 141

Chappell, V.   188 n.2
Chevalier, J.-M.   9
Church, J.   185 n.3
Church-Turing thesis   85
circumstances of evaluation, *see* context, of evaluation
Clark, A.   59
Clement, C. A.   6
Collins, C.   10
compositionality   43–6, 48, 53, 63, 72, 149, 164, 167, 171
concept formation   19
connectionism   47
construction
  abortive   85–6, 89, 146–7
  concept of   14, 16, 66, 78–84, 90–1, 98, 101, 113, 121, 143, 146, 164, 182, 190 nn.2–5
  construction invariants   16, 101, 111, 113–22, 129–30, 134, 153, 155, 166–7, 177, 180, 182, 192 n.6
  definition of   83–9
  indeterminable   146–8, 194 nn.10–11
  indetermined   194 n.10
  operation of   73, 75, 83–6, 90, 97–9, 114, 119, 123, 130, 146, 161, 165–6, 177, 183
  parameters of   83–4, 86–8, 91, 113–15, 117, 119, 130–1, 141, 146, 157, 165, 167, 180
  properties of   89–91, 95, 98–9, 120, 129, 131, 147, 159, 161, 192 n.1
  results of   83–4, 90, 141, 190 nn.5–6
  rules of   14, 16, 75, 79–82, 86–7, 89–91, 93–9, 113, 116–17, 120–2, 129–30, 139–40, 147, 149, 153–5, 161, 163–6, 172, 176–80, 182–3, 187 n.7, 190 n.6
  units of   83, 88, 130
content
  bare bones   33–4
  diagram's   78–9, 86–8, 91, 95, 99, 113, 182, 190 n.6
  distinctively sensory   11, 156
  fleshed-out   33–4
  general   130, 192 n.1, 192 n.3
  iconic   18, 20, 25, 29–30, 37, 58, 86, 129–34, 142, 150, 158, 160–4, 166, 169, 172, 176, 180, 182, 193 n.1, 193 n.5
  indeterminacy of   30–2, 42, 158, 160, 163
  non-propositional   13, 15, 20, 25–6, 48–9, 186 n.5
  particular   62, 89, 130, 192 n.1, 192 n.3, 193 n.5
  phenomenological   11, 156
  propositional content   20, 25, 30, 37–9, 66, 163, 169
context
  communicative   31–3
  context (circumstances) of evaluation   124–5, 128–9, 160, 162
  intensional   30, 48
  modal   30, 159
  propositional context   37
convention   9–10, 103, 118, 136, *see also* iconic convention
Cooper, L. A.   150, 154
Coopmans, C.   21
coordination problem   141
correctness
  accuracy conditions   29, 123, 133–4, 136, 162
  correctness conditions (standards)   24, 70, 79, 87, 124, 133–6, 148, 151, 157
  precision conditions   134, 136, 148
  robustness conditions   135
  trueness   134–6, 148
Crane, T.   13, 29, 35, 37, 143
criterion-less identification   160, 163
Croijmans, I.   10
Cromwell, P.   76
cross-ratio   112–14, 120, 133
Cummins, R.   123, 126, 192 n.1
Currie, G.   185 n.4

Davidson, D.   36, 43, 131, 159
Dawes, A. J.   185 n.3
Defeyter, M. A.   8
Dehaene, S.   189 n.6
DeLoache, J.   186 n.4
demonstrative files   108–9, *see also* object-file
demonstratives   16, 101, 106, 108–10, 122, 182
Denis, M.   25
Dennett, D. C.   13–15, 19, 102–3, 128, 137, 167, 186 n.2

depiction
  experiential theories of 11
  intention-based accounts of 33, 126–7
  make-believe theory of 12
  recognition-based theories of 104–5, 121
  resemblance-based theories of 75, 95–6, 98, 105–6, 124, 182 (*see also* resemblance)
De Regt, H. W. 17, 151–2
descriptionalism 174
description-based semantics 16, 75, *see also* identification by description
De Toffoli, S. 79
Devitt, M. 13, 44, 159
diagrammatic reasoning 9, 21, 91, 186 n.2
diagrams
  approximate definition of the content of 86–7, 129
  token-diagram 89, 98, 130, 191 n.8
  type-diagram 89, 130, 191 n.8
dialetheism 146
Dilworth, J. 24
discursive representations 15, 25, 44, 56, 73
Djordjevic, J. 10
Dokic, J. 107
Dominowski, R. L. 186 n.1
Dorsch, F. 28
drawings 2, 5, 12–13, 33, 38, 68, 87, 89, 94, 109, 123, 126, 150, 161, 167, 175
Dresner, E. 13, 137
Dretske, F. 12, 29, 168
dual-process theories 47
Dummett, M. 14, 35

Earnshaw, R. A. 5
Eaton, M. 24, 29
Eco, U. 44
Egan, D. 32
Elgin, C. Z. 4, 102, 113, 126, 150, 152, 175–6, 193 n.2
Ellis, R. D. 185 n.2
Elpidorou, A. 143
epiphenomenal nature of images 17–18, 20
epistemological challenge 28–30, 63, 149, 151

Ernst, B. 144
Ertz, T.-P. 79
Etchemendy, J. 7
Evans, G. 40–1, 43, 45, 101, 106, 108, 110, 184, 186 n.6, 193 n.4
Evans, J. St B. T. 186 n.1
exemplification 78, 80, 125, 129, 136, 176–7

Falmagne, R. J. 6
Fazekas, P. 189 n.5
Ferguson, E. S. 5, 17
Finke, R. A. 11, 174
Finkelstein, L. 137
Firestone, C. 59, 175
Fish, W. 191 n.2
Fodor, J. 3, 25, 31, 43–4, 46, 48, 50, 112, 164, 166, 168, 194 n.3
Francis, G. K. 145
Frege, G. 13–14, 18, 34–5, 46, 48, 50, 66, 183, 185 n.5, 190 n.3
Frege-Davidson's Argument 30, 34–6, 39, 42–3, 98, 149, 181
French, S. 95–6
French, S. R. D 116
Frigg, R. 6, 129, 191 n.13
Frixione, M. 32

Galton, F. 56
Gardner, H. 17
Gaskin, R. 42
Gauker, C. 19, 25, 57, 60, 62–3, 108, 150–1, 192 n.3
Gendler, T. S. 17
generality constraint 43–6
General Problem Solver (GPS) 46
Gentner, D. 35
Gersel, J. P. 41
gestalt psychology 50–1
Giaquinto, M. 7, 17, 91, 191 n.10
Giardino, V. 7, 9, 22, 79, 103
Gibson, J. J. 111–12, 114
Giere, R. N. 135, 191 n.13
Gilbert, A. N. 10
Giovannelli, A. 107
Goldberg, B. 31
Goldin-Meadow, S. 6, 22
Goldstone, R. L. 58
Gombrich, E. H. 11, 17, 98, 112
Gooding, D. C. 5

Index

Goodman, N.   4, 12, 19, 32, 65, 70, 99, 103, 124–6, 129, 168, 175–6, 184, 191 n.1, 193 n.4
Goodwin, W.   29
Grandin, T.   21
Green, E. J.   44, 59, 108, 114, 121
Greenberg, D. L.   17
Greenberg, G.   9, 29, 103, 123–4, 192 n.1
Gregory, D.   11, 151, 156, 186 n.5
Gregory, R. L.   144
Grzankowski, A.   13, 37
Guillot, A.   10
Gurr, C.   22

Hacker, P.   18, 64–5, 69
Halpern, A. R.   10
Harshman, R.   188
Haugeland, J.   12, 33, 168
Hawthorne, J.   41
Heck, R. G.   29, 35, 43
Hegarty, M.   21, 79
Heuer, F.   174
Hintikka, J.   37
Hinton, G.   174
Holyoak, K. J.   186 n.1, 188 n.1
homomorphism   191 n.12, 193 n.7
Hookway, C.   9
Hope, V.   106–7
Hopkins, R.   11, 105, 127
Hubbard, T. L.   8, 10
Hume, D.   47, 58, 187 n.8, 188 n.2
Hummel, J. E.   107, 192 n.5
Husserl, E.   131, 193 n.2
Hyman, J.   80, 99, 105, 124, 193 n.4

iconic convention   120–2, 124–5, 129–33, 142, 194 n.8
iconic denotation   126–9, 131–2
iconic difference   3
identification by description   106, 110
image (icon)
    auditory   5, 10, 12–13, 139–40, 165
    emotional   10
    external (pictures)   9, 12, 185 n.4
    gustatory   10, 164, 170
    haptic   10
    iconic representation   3, 8–12, 21, 42, 44, 99, 121, 164, 168–73, 178, 180, 183, 192 n.3
    impossible   123–4, 142–8, 194 n.11
    literary (metaphor)   10, 12
    measurement-theoretic account of   137–42
    olfactory   10, 12, 139–40
    tactile   10
imageless thought controversy   18
image-propositions   32
imagistic cognition   62–3
imagistic theory of thought   13, 19, 21–3, 25–6, 55, 57–9, 73, 149, 158, 175, 183, 189 n.3
impossible fork   124, 143, 145–7
indexicals   132, 136, 152, 154, 165
indices   9–10, 108, 138
informative richness   169
Ingarden, R.   193 n.2
Inkpin, A.   112
instrumental nature of images   17–18, 20, 22, 26, 28, 181
intentionality   39–40
Intons-Peterson, M. J.   7
irreducibility thesis   15, 19–20, 23, 25–6
Irvine, E.   185 n.2
Isaac, A. M. C.   117
Ishai, A.   174
isomorphism   128, 191 n.12
Ison, M. J.   116
Ittelson, W. H.   174
Izard, V.   189 n.6

Jackson, F.   45
Jäkel, F.   51
Jakubowski, K.   10
James, M.   111
Johnson, K.   194 n.2
Johnson, M.   185 n.2
Johnson-Laird, P. N.   6, 22, 188 n.1

Kamermans, K. L.   174
Kan, I. P.   58
Kant, I.   27, 66, 70, 184, 187 n.11, 190 n.4
Kaplan, D.   33, 37, 40, 110, 128, 155
Kasem, A.   188 n.2
Kaufmann, G.   19
Kennedy, J. M.   144
Keogh, R.   177, 185 n.3
Khalifa, K.   152
Kiefer, M.   58
Kind, A.   28
Kinzler, K. D.   189 n.3

Kirsh, D. 21, 185 n.4
Kitcher, P. 21, 29, 169
Kjørup, S. 24
Klatzky, R. L. 10
Knauff, M. 6
knots
    knot diagrams 16, 75–9, 86, 120, 182, 189 n.1
    knot invariants 77
    knot theory 75–9
knowledge
    discriminating 40
    imagistic 24, 149–58, 180, 182
    knowledge-by-acquaintance 151
    non-propositional 80, 132, 150–1
    phenomenal knowledge 151
    propositional knowledge 24, 28–9, 79, 86, 110, 150–1, 155, 183
Kobayashi, M. 10
Kosslyn S. M. 17, 56, 173–5
Kosso, P. 152
Kozak P. 22, 194 n.4
Kozhevnikov, M. 21, 188 n.1
Krantz, D. 193 n.7
Kripke, S. A. 31, 126
Kulpa, Z. 7, 143
Kulvicki, J. 3–4, 10, 12, 24–5, 33, 37, 93, 116, 144, 168, 171, 186 n.3
Kvanvig, J. L. 152

Lakoff, G. 185 n.2
Langacker, R. W. 185 n.2
Langland-Hassan, P. 24
Larkin, J. H. 21–3, 28, 169
Latour, B. 21
Laurence, S. 43
Lehrer, K. 80
Lemon, O. 22
Lewis, D. 168, 189 n.8
Lieberman, K. 56
Lindsay, R. K. 22
Lipton, P. 151
List, C. J. 32, 187 n.1
Liu, Z. 110
Locke, J. 50–1, 57–60, 66, 73, 175, 188 nn.2–3
Loewenstein, J. 35
logical form 5, 30, 34–7, 39, 42, 136, 149–50, 158, 166, 169, 172, 181
logic of showing 3

Lombardii, A. 32
Lopes, D. M. 4, 10, 104–5, 111–12, 127
Lowe, D. G. 116
Lowe, E. J. 185 n.2, 188 n.2
Lupyan, G. 35

Macbeth, D. 7
McCarthy, R. A. 109
McDowell, J. 191 n.2
McGinn, C. 168, 174
Machery, E. 58, 189 n.6
MacKisack, M. 173
McNeill, D. 6
Maglio, P. 185 n.4
Mahon, B. Z. 189 n.6
Maley, C. J. 168
Malinas, G. 36–7
Mancosu, P. 7
Mandelbaum, E. 49
Manders, K. 7
Manley, D. 41
Marcus, R. B. 13, 137
Margolis, E. 43
Martin, A. 58, 189 n.4
Mast, F. W. 174
mathematical proof 91, 190 n.2, 191 n.10
Matthen, M. 37, 45, 110
Matthews, R. J. 13, 137
Maynard, P. 137, 185 n.4
measurement
    concept of 127, 131, 134, 137–42, 156
    devices 14–15, 26, 75, 127–8, 137, 142, 148–50, 152–3, 156–8, 170, 176, 182–3, 193 n.8
    indications 156
    invariants 118
    outcomes 135, 138–40, 156
    predicate 158–9
    representational theory of measurement (RTM) 193 n.7
mental imagery
    concept of 11, 19, 173–80, 184
    dilemma of 175
    mental imagery debate 3, 7, 173–4, 186 n.2
    photographic model of 175
    weak perception theory of 174–5, 178

mental images
  concept of   12, 173–80, 185 nn.3–4
  constructive account of   176, 179
  experiential account of   176
  unconscious   11, 176–7
mental representation format
  analog   168–70, 182
  concept of   168
  digital   168–70, 183
  domain-generality of   170, 172–3, 183
  domain-specificity of   170, 172–3, 178, 180, 183
  hierarchical information architecture of   171–3, 183
  holistic nature of   171–3, 180, 183
  non-hierarchical information architecture of   171–3, 183
mental rotation tasks   21, 155, 167, 175, 177, 179
Merricks, T.   13
Mersch, D.   3, 36
metaphysical challenge   43–6, 53, 63, 164, 167
Metzler, J.   17, 56, 154, 173
Meynell, L.   17
Michaelian K.   192 n.3
Mitchell, W. J. T.   3
Miyake, A.   56
Molitor, S.   8
Montague, M.   13
Morgan, M. S.   138
Morrison, M.   138
Morrison, R. G.   186 n.1
Mortensen, C.   142, 145–6
Mößner, N.   5, 25, 63, 129, 151
Mullarkey, J.   19
Müller, A.   3
Mumma, J.   7
Murez, M.   108

Nail, T.   10
Nanay, B.   4, 10, 174, 178, 185 n.3, 189 n.5, 195 n.8
Narens, L.   137
nativism   57, 189 n.3
natural generativity   104, 121, 166
Neander, K.   105
Necker cube   91, 143, 163
Nelsen, R. B.   7

neo-empiricist consensus   58
neo-Lockean approach   15, 56–7, 60, 63, 68, 175
Nersessian, N.   5
Newall, M.   11, 105
Newell, A.   46, 50
Newton, N.   10
Nguyen, J.   129
notation
  concept of   22
  Conway notation   77
  Dowker notation   77
Novitz, D.   3, 105
Nunberg, G.   38
Nyíri, J. C.   31, 185 n.2

Oberlander, J.   191 n.10
object-file   108–9
O'Brien, D. P.   12
operational approach   15, 56, 67–70, 74–5, 181, 189 n.7
optical information   111–12, 120
Otis, L.   188 n.1
Overgaard, S.   106

Pagin, P.   44
Paivio, A.   17, 56, 173, 188 n.1
Peacocke, C.   11–12, 43–4, 107–8, 168–9
Pearson, J.   56, 174, 177, 185 n.3, 195 n.6
Pecher, D.   58, 189 n.4
Peirce, C. S.   4, 9–10, 91–2, 184, 187 n.11, 190 n.4, 191 n.8
perception, see also thought, thoughts-perception border
  cognitive impenetrability of   175
  object perception   106–7
  perceptual illusions   144
  perceptual invariants   111, 114
  perceptual recognition   106–8, 111, 116, 179, 191 n.2
  that-perception   106–8, 112
Perini, L.   29
Pero, F.   191 n.12
Perry, J.   152
Peterson, M. A.   174
pictorial fallacy   36–9, 90, 159, 186 n.4
pictorialism   173
picture principle   44, 164
picture superiority effect   8
Pietarinen, A.-V.   9, 186 n.2, 191 n.12

Podro, M. 11
predication (predicative function) 39, 41–2, 48, 158–62, 170–1, 186 n.4, 193 n.4
Price, H. H. 4, 19, 69, 187 n.1
Priest, G. 143–6
principle of charity 131
Prinz, J. J. 57–8, 60–2
private language argument 31, 65
Proclus 81, 190 n.2
projective geometry 82, 112–14, 116, 120
properties
    first-order 92, 94, 96, 117
    second-order 93–5, 105, 117
propositional attitude 1, 12–13, 29, 46, 51, 107
propositionalism (propositional theory of thought) 46, 48–9
proxytypes 60–1, 63
Pulvermüller, F. 58
Putnam, H. 93, 124
Pylyshyn, Z. W. 3, 7–8, 18, 25, 43, 45, 48, 55, 59, 108, 112, 154, 166, 168, 173

Quilty-Dunn, J. 44, 49, 108, 171
Quine, W. V. O. 38, 40, 69, 194 n.2
Quiroga, R. Q. 116

Raftopoulos, A. 59, 175
Ragni, M. 6
rationality 36, 46, 48–9, 65, 150
Rayo, A. 83
Recanati, F. 108
Received View 15, 27, 46–9, 52–3, 181, 186 n.1, 187 n.8
recognition (recognition-based identification) 16, 75, 80, 101–24, 129–31, 152, 154, 156, 166, 179–80, 182–3, 191 n.2, 192 nn.4–5
reference
    causal models of 126
    deferred 38
    direct 9, 12, 105, 108, 111, 116, 122, 130–2, 163
    iconic 124–6, 134, 136, 148, 182
    non-rigid 133, 160
    rigid 133, 160
    two-dimensional model of iconic 16, 75, 123–33, 147, 182

referent 16, 37, 41, 89, 123–5, 129–36, 139, 142, 148–9, 160–2, 164–7, 182, 192 n.1
Reidemeister moves 76–9, 85, 87
Reidemeister's theorem 76
Reisberg, D. 7, 174
Rescorla, M. 170, 187 n.9
resemblance 9, 32–3, 36, 92–9, 105, 122, 124, 142, 157–8, 161, 182, 191 n.11, *see also* depiction, resemblance-based theories of
Reutersvaard-Penrose triangle 143–6, 194 n.9
Rey, G. 13
Rheingold, H. R. 5
Richardson, J. T. E. 187
Richtmeyer, U. 31
riddle of style 98, 191 n.13
Riemann, B. 82, 114
Rocke, A. J. 188 n.1
Rogers, Y. 8, 174
Rollins, M. 19
Roser, A. 32
round square 147
Rozemond, M. 41
rules, *see also* construction, rules of
    rules of interpretation 80–1, 87
    rules of production 80–1, 87
Runhardt, L. 137
Russell, B. 19, 32, 40, 55, 187 n.1
Russell's Principle 40–2, 108
Ryle, G. 5, 18, 27, 64–6, 69, 71, 179, 187 n.10

Sainsbury, R. M. 35
Salis, F. 6
Salje, L. 194 n.2
sample 23, 42, 125–6, 135–6, 139, 157, 172
Sartre, J.-P. 174
Sartwell, C. 104
Scaife, M. 8, 174
Scarry, E. 10
Schier, F. 11, 99, 104–5, 124
Scholl, B. J. 59, 175
Schooler, J. W. 50
Schöttler, T. 106
Sebeok, T. A. 10
seeing-in 96, 124, 131, 192 n.6
Sellars, W. 1, 3, 13–14, 19

semantical challenge 39–43, 149, 158, 186 n.5
Shabel, L. 190 n.4
Shah, P. 56
Shaver, P. 6
Shepard, R. N. 17, 19, 56, 150, 154, 173, 188 n.1
Sheredos, B. 5, 185 n.4
Shimojima, A. 22, 186 n.2
Short, T. S. 9
Siewert, C. 107
Simon, H. A. 21–3, 28, 46, 50, 169
Skinner, B. F. 51
Slezak, P. 2, 64, 174–5
Sloman, A. 5, 28
Sober, E. 44
Soles, D. 188 n.2
Solomon, K. O. 43
Solt, K. 29
Sorensen, R. 142
space
    informational 83, 85–6, 90, 99, 113, 117, 121, 129
    logical 14, 51–2, 83, 86, 105, 114, 127, 129, 131–2, 152–4, 157, 165, 176, 180, 182, 191 n.7
    perceptual similarity 62–3, 108
    physical 14, 51–2, 83, 86, 105, 127, 131–3, 152–4, 157, 165, 180, 182, 191 n.7
spatial representations 186 n.2, 194 n.11
Spelke, E. S. 189 n.3
Squires, R. 104
standard metre, 118, 133, 135, 140, 162, 178, *see also* measurement
Stanley, J. 151
Stegmüller, W. 66
Stenning, K. 12, 22, 191 n.10
Sternberg, R. J. 6
Stich, S. P. 35
Stjernfelt, F. 9, 22, 186 n.2
Strawson, P. F. 109
structural similarity 93–6, 98, 116
Suárez, M. 32, 95, 191 n.12
Swoyer, C. 17, 93, 131, 135, 137, 186 n.2
symbols (symbolic representations) 6, 9–10, 18, 22, 25, 31, 69, 72, 90, 104–5, 121, 129, 187 n.7, 190 n.4

systematicity 15–16, 31, 43–6, 48, 53, 63, 109, 112, 137–8, 149, 164–7, 180–2, 186 n.1, 186 n.6

Tal, E. 135, 141, 156
target 16, 83, 87, 89, 108–11, 114–16, 123–33, 135–6, 141–2, 146–7, 149, 156, 160–3, 182, 191 n.13, 192 n.1, 193 n.2, 193 nn.4–5, 194 n.11
Tarski, A. 31
Teller, P. 135
Terrone, E. 111, 127, 129
Teufel, C. 189 n.5
Thagard, P. 10, 17, 21, 188 n.1
Thielen, J. 178
thinking
    animal 19
    imagistic 2, 4, 11–15, 18, 20, 23–8, 34, 36, 42–3, 48–9, 51–3, 55–60, 62–3, 68, 70, 73, 75, 149–81, 187 n.1, 188 n.1
    individual differences in 188 n.1
    language-like 2, 5, 12–13, 18, 44, 46, 62, 173, 181
    measurement-theoretic account of 13–15, 149, 179–80, 183
    metaphors of 49–52
    propositional 12–13, 25, 182
    thinking styles 188 n.1
Thomas, N. J. T. 11, 185 n.2
thought
    bearer of 15, 20, 26–8, 31–2, 45, 53, 56, 65, 67, 69–70, 164
    imagistic 2, 4, 11, 15, 18, 20, 23–7, 42, 51, 56–60, 63, 68, 70, 73, 75, 149, 181, 187 n.1
    thought acquisition 47, 57
    thoughts-perception border 59–64, 73, 179
Tichy, P. 82, 190 n.3
Traditional View 15, 39, 43, 73, 75, 140, 142, 148–50, 157–8, 161, 164, 175, 181–2
translatability thesis 15, 23–6, 49
Troscianko, E. T. 10
truth
    truth-bearers 28–9, 31, 34–5, 150
    truth-conditions 28, 31, 36, 38, 135, 185 n.5
    truth-makers 29

T-schema 135, 172
Tufte, E. R. 5
tunnel effect 108, 119
Tversky, B. 21, 165, 187 n.7
Twardowski, K. 66, 171
Tye, M. 29, 151

understanding 151–8, 165–6

Vallée-Tourangeau, F. 6
van Fraassen, B. C. 4, 96, 117, 132, 138, 141, 153
Varzi, A. 21, 29, 164, 169
Vernazzani, A. 191 n.2
Viera, G. 10
visual impedance effect 6
Voltolini, A. 11, 144
von Neumann, J. 170

Walton, K. L. 12, 99, 105, 124
Ware, C. 22
Warrington, E. K. 109, 111
Watson, D. 5
Watson, J. B. 18
Westerhoff, J. 36
Westerståhl, D. 44

White, A. R. 188 n.2
Wiesing, L. 193 n.2
Willats, J. 103
Williamson, T. 28, 150–1, 157
Winn, W. 8
Wittgenstein, L. 18, 30–4, 45, 64–6, 162–3, 184
Wittgenstein-Ryle's sceptical argument 15, 56, 64–7, 73–4, 90
Wittgenstein's argument from content indeterminacy 30–4, 39, 42, 45, 96, 98, 149–50, 161, 163, 181, 187 n.1
Wollheim, R. 11, 32, 44, 96, 104–5, 119, 124, 131, 187 n.1

Yolton, J. W. 188 n.2
Yoo, S.-S. 10
Yule, P. 12

Zacks, J. M. 185 n.1
Zagzebski, L. 151
Zatorre, R. J. 10
Zeimbekis, J. 29, 59, 110
Zeman, A. Z. J. 11, 177, 185 n.3
Zhang, J. 21
Zhao, F. 2

www.ingramcontent.com/pod-product-compliance
Lightning Source LLC
Chambersburg PA
CBHW071830300426
44116CB00009B/1494